Environments Through Time

A Laboratory Manual in the Interpretation of Ancient Sediments and Organisms

Robert L. Anstey
Michigan State University
East Lansing, Michigan

Terry L. Chase
Permian Basin Petroleum Museum
Midland, Texas

Burgess Publishing Company
Minneapolis, Minnesota

Printed in the United States of America
SBN 8087-0117-7

3 4 5 6 7 8 9 0

Preface

The earth's past has been a complex interaction of physical and biological processes whose end result is the present-day environment. The best record of the ancient environments of the earth's surface is provided by the constituent particles of sedimentary rocks, of both organic and inorganic origin, and the patterns made up by those rock types, both in space and time. The history of the hydrosphere, atmosphere, and biosphere can be read in the successive layers of sedimentary rock that form the uppermost part of the earth's crust. The exercises in this manual have been designed to enable the student to make the types of interpretations upon which the physical and biological history of the earth is based.

Much of the work in this manual depends heavily upon the concept of *uniformitarianism*, namely that present-day environmental processes are the key to the interpretation of ancient rocks. We owe this concept to James Hutton (1795), the father of modern geology. Uniformitarianism remains geology's major contribution to the philosophy of science. In examining ancient rocks, the student should try to assess whether or not environmental processes operated in the past with intensities similar to those of the present day, if indeed they operated at all. Although uniformitarianism is still a useful introduction to geologic thinking, it provides only a first approximation to what was real in the geologic past. Future work in geology will undoubtedly show increasing deviation from the uniformitarian model.

In the sequence of exercises that follows, the student builds from the small to the large in developing the interpretive process. The first exercise deals with the size and shape of sediment grains, the smallest objects of study. He subsequently proceeds to small three-dimensional grain aggregates, which are called *sedimentary structures*. The next step is to examine the geometry of the sediment bodies that result from the dynamics of the major complexes of depositional environments. Lastly he will examine regional sedimentation across a large portion of the global surface through time. These exercises include the effects of both organic and inorganic processes.

A prevailing theme in the study of historical geology is the concept of time. The duration of time with which geology deals sets it apart from most of the other sciences. The following chapters emphasize the role of time in generating lithologic and paleobiologic patterns, as well as how to interpret the time relationships of the different parts of the rock record.

Note to Students

Worksheets and tracing paper for many of the exercises will be found at the back of the book.

Acknowledgments

We acknowledge the invaluable help of the many students and teaching assistants who have suffered through the preliminary versions of this manual. We especially thank Dr. Mary W. Davis for her help in generating Chapters 2 and 13. We thank Mr. Carl Steinfurth for his help with Chapter 11.

Contents

Chapter One

Sedimentology
Size and Shape of Detrital Grains

Objectives

1. To understand the simple sedimentary processes of weathering, transportation, and deposition.
2. To determine the effects of different amounts of kinetic energy on the roundness, size, and sorting of sediment grains.
3. To compare and contrast the roundness, size, and sorting of samples taken from different sedimentary environments.

Important Terms

Detrital—Derived by the weathering and transportation of a preexisting rock; usually composed of silicate minerals.

Roundness—The absence of angular corners and edges from a sediment grain.

Sorting—The absence of grain size variation within a sediment.

Discussion

Sedimentation involves the removal of a rock grain from its parent material (weathering), its transportation by a fluid medium (air or water), and its final settling out of that fluid (deposition). According to a physical relationship known as Stokes' Law, larger particles will settle out of quiet water, and smaller particles will remain in suspension. The higher the water velocity or degree of water agitation, the larger the particle that can stay in suspension. Grain roundness depends upon the number of times that grains come into mutual contact. Different degrees of grain roundness are shown in Figure 1.1. Roundness could reflect either the degree of water agitation or the length of time that the grain has remained in suspension.

Sorting is the inverse of size variation of grains within a sediment. Samples with little variation are well sorted, whereas those with considerable variation are poorly sorted. To observe the size variation seen in different degrees of sorting, mark the grains in each of the four sorting images shown in Figure 1.2 which have the largest and smallest diameters, respectively. What can cause size

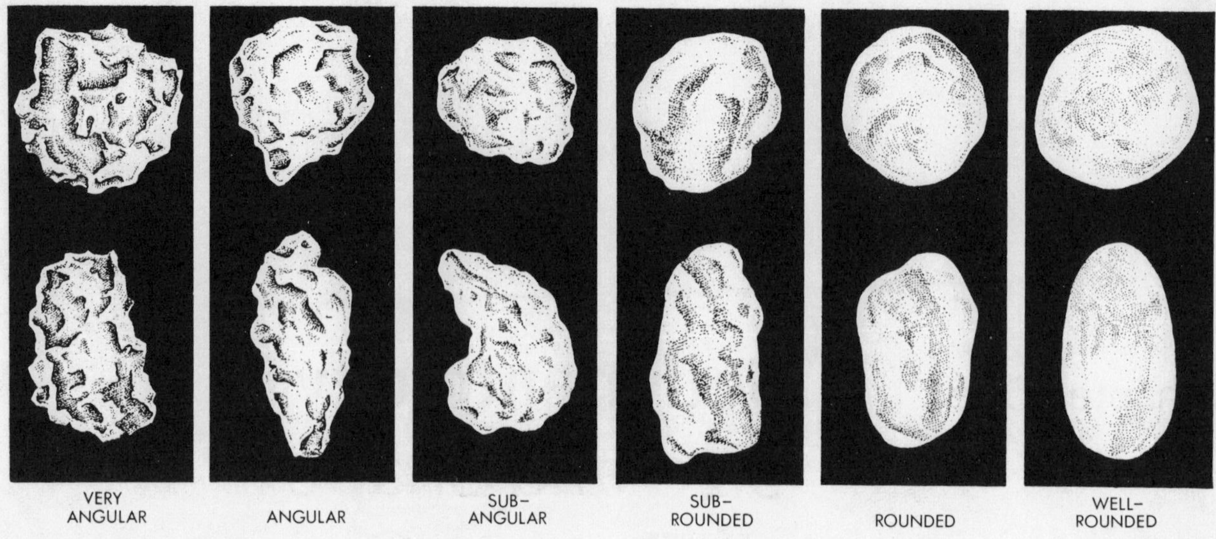

Figure I.1 Grain roundness images

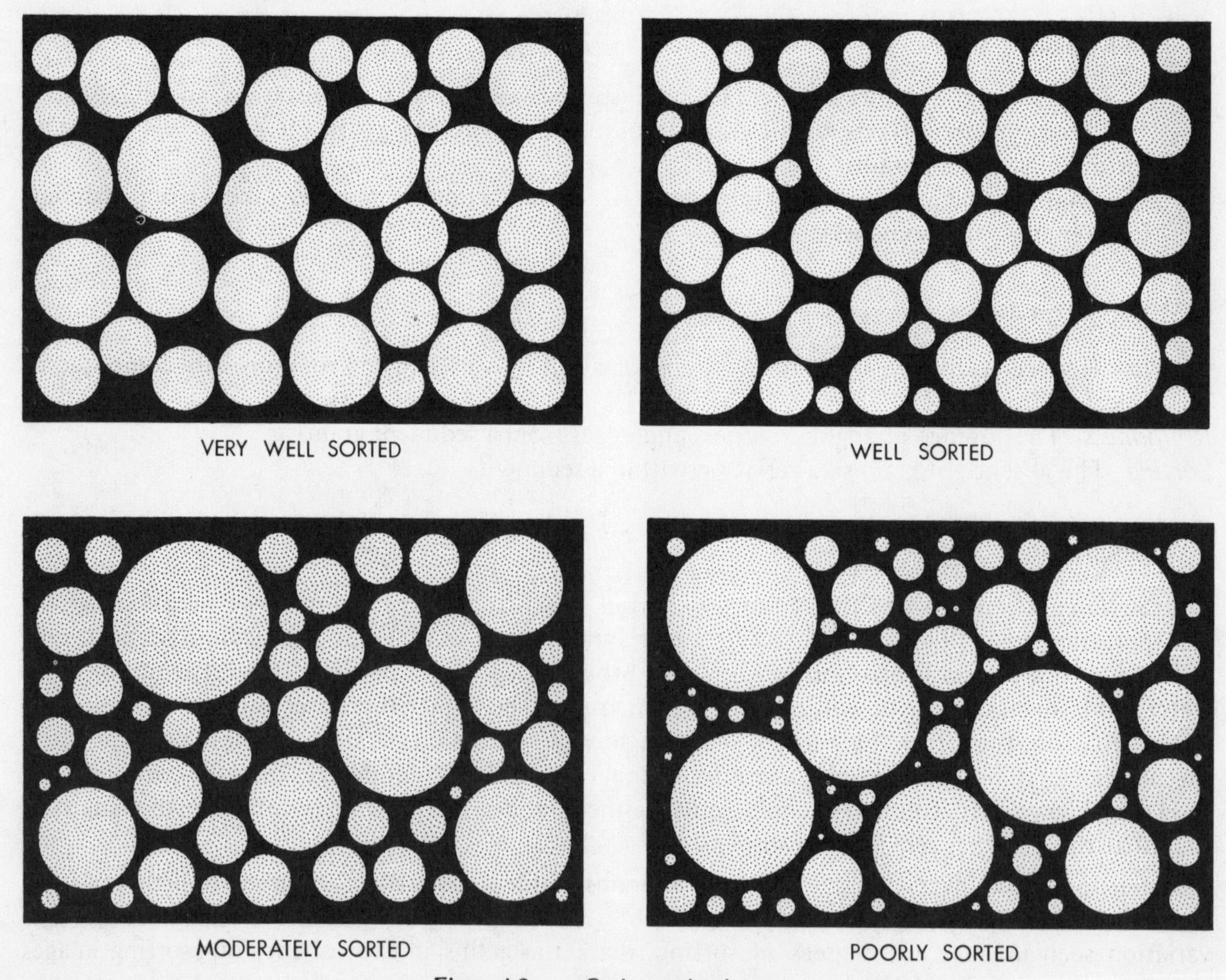

Figure I.2 Grain sorting images

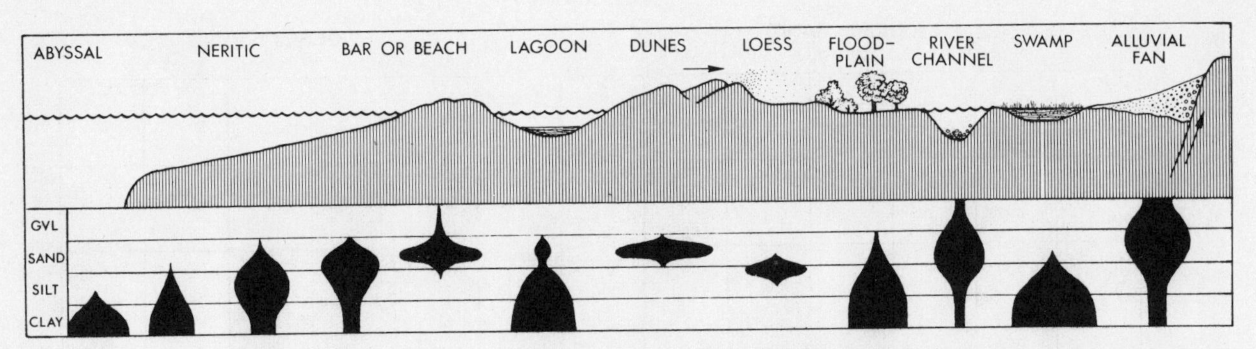

Figure I.3 Grain size distributions in major depositional environments

variation within a sediment? One possibility is that the sediment either has not been transported far or has not been extensively worked by water, giving it a similar mixture of grain sizes to the untransported, unworked soils weathering from the parent material. Very poor sorting is found in alluvial fans and glacial tills, in the former because of lack of transport, and in the latter because of no working by water. Another possibility is that a sudden loss of kinetic energy by the transport medium could cause a wide range of sediment sizes to be deposited simultaneously. This happens when a stream experiences a sharp change in gradient (as when flowing over an escarpment), or when it loses velocity flowing around a meander loop in a river, dropping its sediment in deposits on the inside margin of the curve. Conversely, a relatively constant amount or high level of kinetic energy, as in waves breaking on a beach, could result in a well-sorted sediment.

Figure 1.3 shows the grain size distributions for twelve major depositional environments. As an exercise, rank these both in order of average grain size and sorting. What is the cause of the observed sorting in each environment? Grain size distributions are frequently shown as a bar graph (Figure 1.4) in which the height of each bar equals the percentage of the total sediment (by weight) that each particular size interval represents.

Exercise

Sediment samples from different depositional environments will be provided in the laboratory.

1. Before beginning the analysis, write down the predicted roundness, grain size, and sorting for each sediment, based upon your knowledge of the sedimentary processes that characterize each environment.
2. Examine each sediment under a binocular microscope to test for roundness. Using Figure 1.1, determine the roundness of 10 randomly selected grains. What is the *dominant* roundness of each sample?
3. Use the sediment splitter to obtain about 40 grams of each sample. Separate the sediment into size fractions using an appropriate collection of nested sieves. Your instructor will explain the technical details of sieve analysis. Although sieving does not yield precise results, it will give a rough approximation of the sediment size distribution. Weigh each size fraction and present the results as a bar graph, similar to those in Figure 1.4.
4. Compare your results to your predictions, and discuss possible reasons for any disagreement that you might have found.

Test Questions

1. Does any correlation exist between the size, sorting, and roundness observed within a sediment? What could cause this relationship to exist?

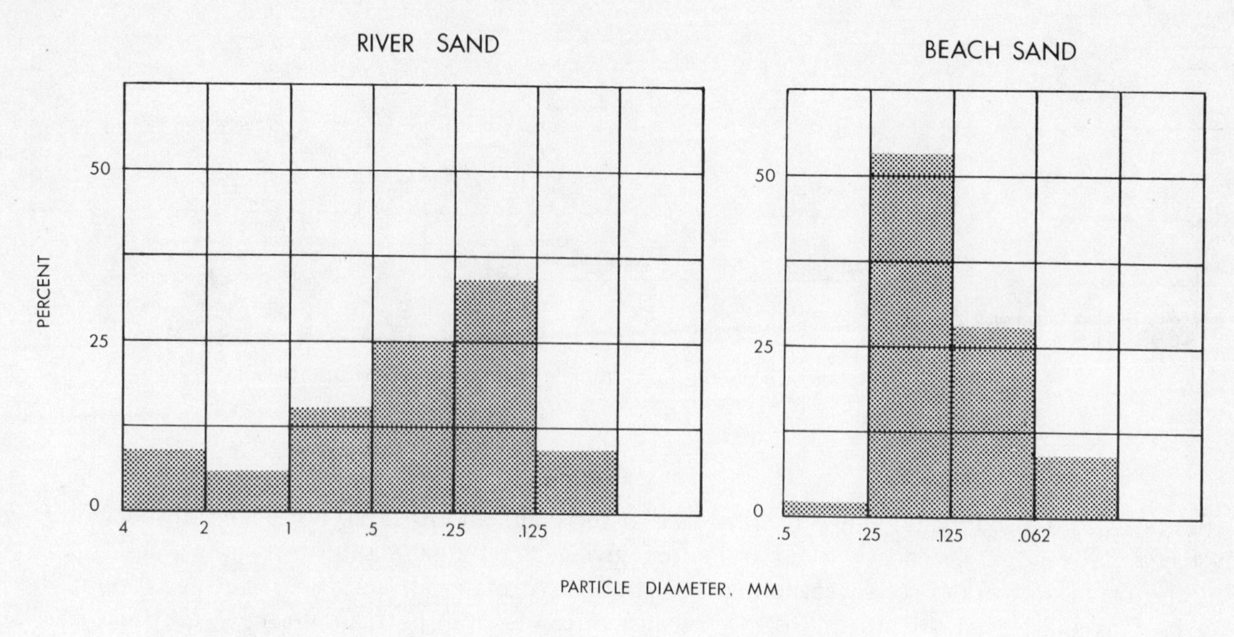

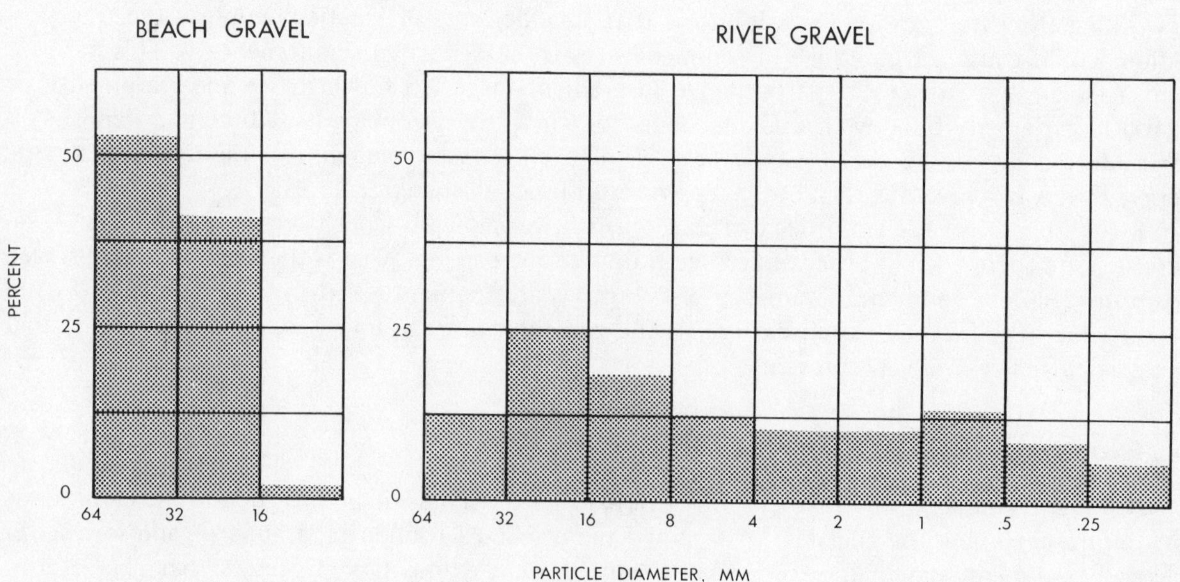

Figure I.4 Grain size distributions of beach and river sediments

2. What could cause grains of different size to have different degrees of roundness within the same sediment?
3. Is there any correlation between the age of a sediment particle (time since erosion) and its size? Could different particles within the same sediment be of different ages?
4. What could cause the presence of well-rounded grains in a poorly sorted sediment?
5. Can all sedimentary environments be differentiated by the size and shape of their grains? Which environments are distinctive, and which are not?

Background Reading

Folk, R. L., 1965, Petrology of Sedimentary Rocks, Hemphill's, Austin, Texas, p. 1-52.

Krumbein, W. C., and Sloss, L. L., 1963, Stratigraphy and Sedimentation, 2nd ed., W. H. Freeman and Co., San Francisco, p. 96-114.

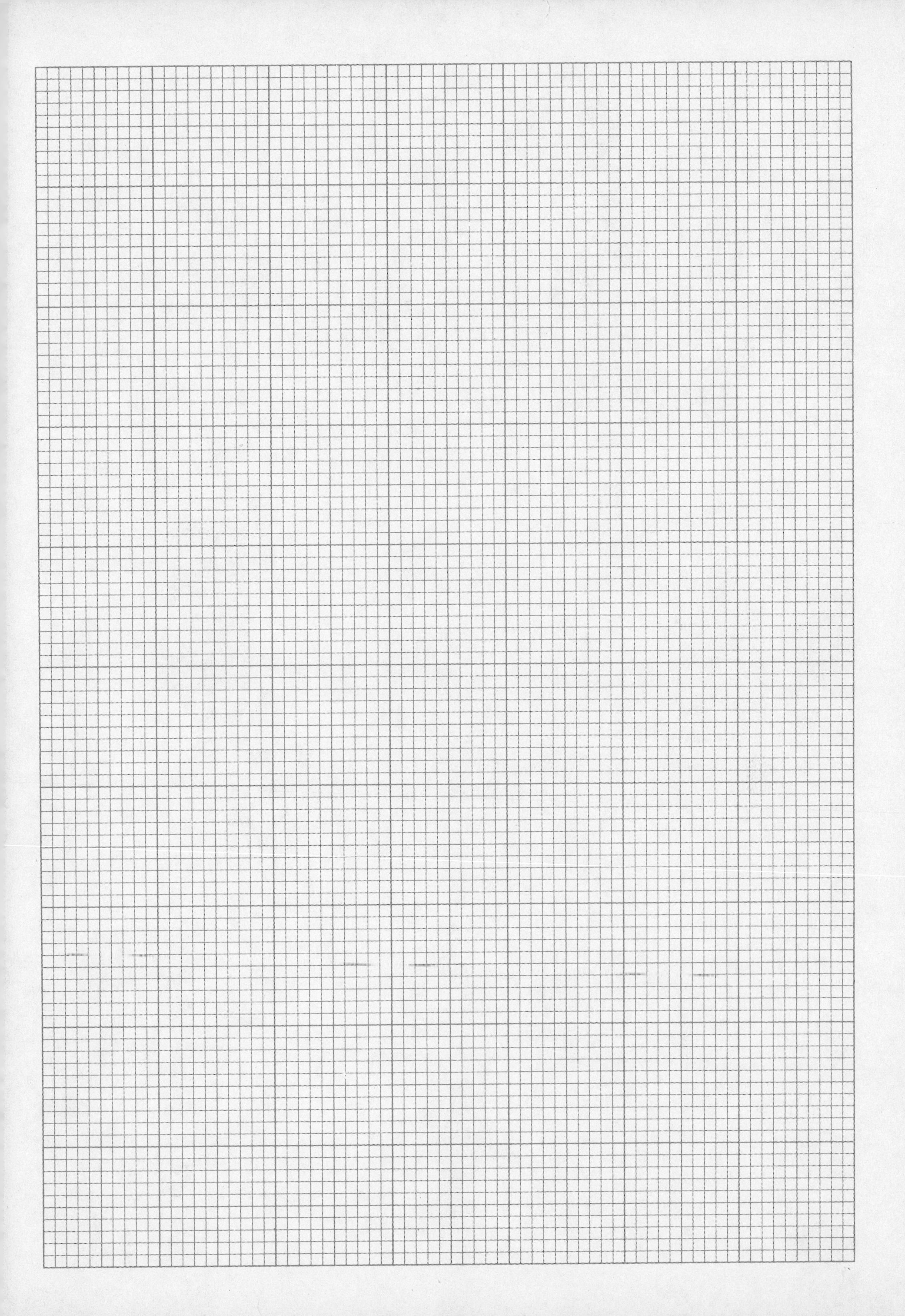

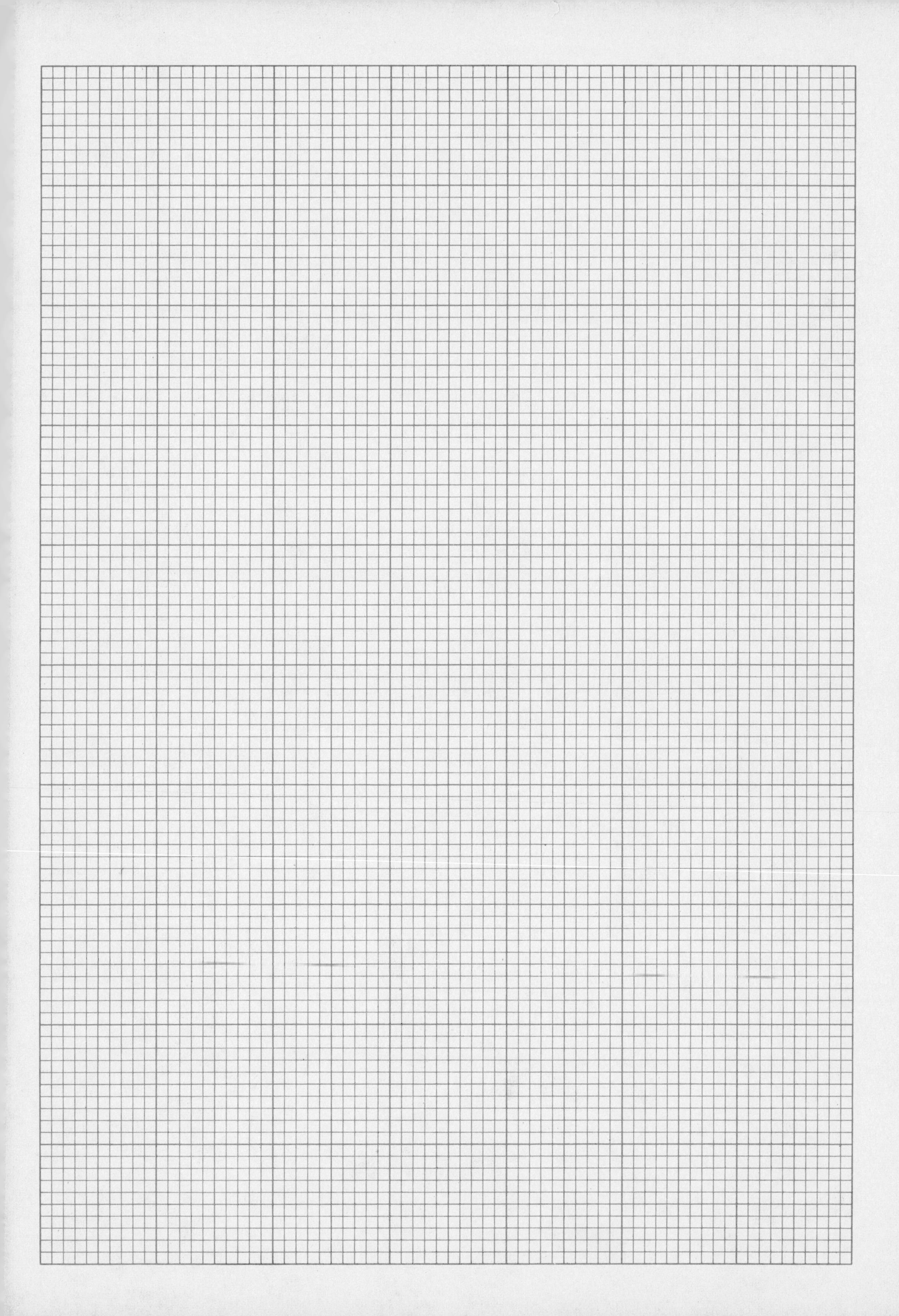

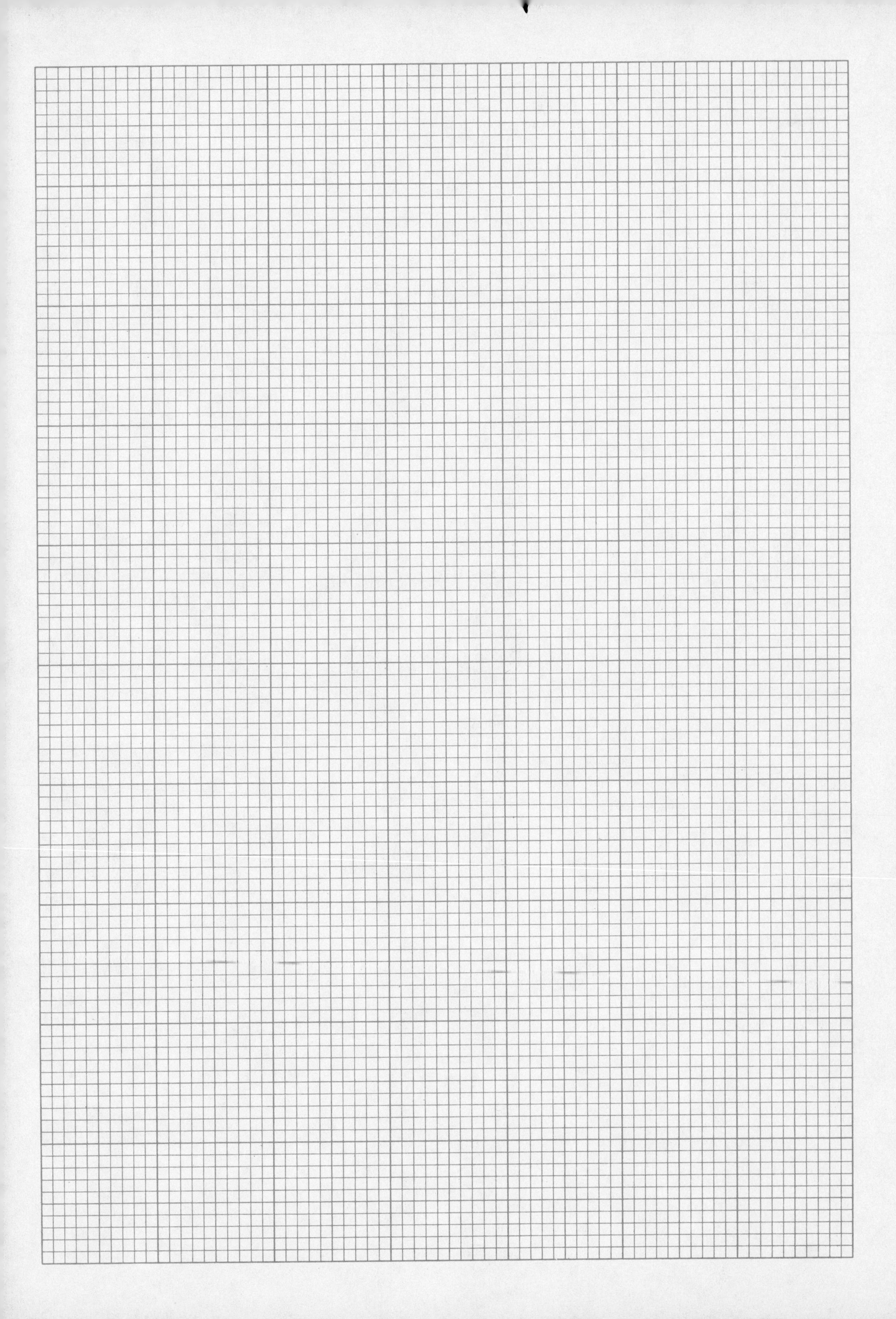

Chapter Two

Sedimentary Petrology Origin of Detrital Rock Types

Objectives

1. To recognize the grain and aggregate properties of the major types of detrital sedimentary rocks.
2. To interpret the origins of these rocks in terms of source materials, transport, and depositional environment.

Important Terms

Detrital—Derived by the weathering and transportation of a preexisting rock.
Chemical—Derived by chemical or biochemical processes in the place of deposition and not transported.
Tectonics—Crustal readjustment causing rock deformation through faulting or folding.

Major Categories of Sedimentary Rocks:

Class	*Common Rock Types*	*Source of Grains*
Detrital	Mudstone Siltstone Sandstone Conglomerate	Erosion of preexisting igneous, metamorphic, and/or sedimentary rocks; transported to depositional site.
Chemical	Limestone Dolomite Evaporites Coal	Plant and animal skeletons or direct precipitation from water.
Mixed	Shaley Limestone Organic Shale Sandy Dolomite Limy Siltstone	Material from both sources

Detrital sedimentation requires that streams and rivers transport large amounts of eroded material to the depositional site. Hence detrital rocks imply the existence of an actively eroding source area, probably the result of contemporaneous tectonic uplift. Chemical sedimentation requires that the supply of detrital material be small, either through tectonic quiescence or great distance from a river mouth. All three major rock classes may form in close proximity, however, in deltas, shorelines, and lakes.

Detrital Minerals

The types of minerals in a detrital rock depend upon both the source rock's composition and the length of time that the eroded minerals have been subject to further decomposition. The stability of a mineral at the earth's surface is crudely the inverse of its original temperature of crystallization. Stability roughly follows the spectrum below:

Stability	Ferromagnesian Silicates	Aluminosilicates
Lowest	Olivine	Calcium Plagioclase
↓	Pyroxene	Sodium Plagioclase
	Amphibole	Potassium Feldspar
	Biotite	Muscovite
	Limonite/Hematite	Quartz
Highest	Clay minerals	Clay minerals

The most common minerals in detrital rocks, therefore, should be clay minerals, quartz, limonite, hematite, and micas. Feldspars and ferromagnesian silicates should be rare except in sediments that have neither been transported far nor extensively water-worked, such as in alluvial fan deposits. This explains why most detrital rocks are either sandstones or mudstones, commonly having yellow or red iron staining.

The mineral composition of detrital rocks can also reflect the depositional environment, as well as the source area. A stream-deposited sandstone may contain feldspars, micas, and rock fragments, whereas beach sands consist almost entirely of quartz. Why do these two different environments produce different mineral compositions in their sediments?

Grain Properties

Grain roundness has been treated in Chapter 1 (Figure 1.1). Grain size reflects the velocity of the transporting wind or water. The water on a dirt road after a rainstorm moves clay and silt; the water on a beach moves sand and pebbles. Thus sandstones and conglomerates accumulated in beaches and streams; shales and mudstones were deposited in swamps, quiet bays, or lagoons (Figure 1.3). Mixtures of size classes or intermediate size classes reflect intermediate conditions or the gradual encroachment of one environment on another. Detrital rocks are classified according to their most common grain size:

Grain Diameter	Rock Name
2 mm to 64 mm	Conglomerate
1/16 mm to 2 mm	Sandstone
1/256 mm to 1/16 mm	Siltstone
Less than 1/256 mm	Claystone or Shale

Aggregate Properties

Sorting has been discussed in Chapter 1 (Figures 1.2, 1.3, and 1.4). Because sorting reflects water velocity, which is usually fairly constant over long periods of time, many sediments are moderately well sorted. Poorly sorted sediments reflect great variations in water velocity, such as those that take place during floods or storms.

Much of the volume of many rocks comes from the organic tissues and skeletons of plants and animals. Considerable organic material is present in fine-grained sediments, and imparts a black color to the rock. This material consists of free carbon and hydrocarbons, and is very common in shales. Why is this material less abundant in sandstones?

Commonly organic material can be recognized as parts of formerly living plants and animals, which are termed *fossils*. Fossils constitute a considerable percentage of the volume of detrital rocks (up to 50 per cent), and the bulk of most carbonate rocks. Knowledge of the life habits of analogous living organisms may allow interpretation of the conditions in which the fossiliferous rocks were formed.

Depositional Environments

The *depositional environment* is the area over which a particular rock type accumulates. Three major depositional complexes can account for the vast majority of all detrital rocks:

1. Shoreline and coastal environments, dominated by longshore currents;
2. Stream, river, and delta systems, including lakes; and
3. Alluvial fans, wedge-shaped deposits at sharp changes in stream gradients, most commonly found along fault escarpments.

Padre Island along the Texas coast (Figure 2.1) is a classic example of coastal sedimentation. The barrier island consists of beach and dune sand, while mud and silt accumulate in the bay, lagoon, and marine shelf environments.

The Birdsfoot Delta of the Mississippi River provides a standard example of river (*fluvial*) and deltaic sedimentation (Figures 2.2., 2.3). In deltas many types of detrital rocks accumulate in close proximity, and often intergrade. Sands are deposited on the inside curves of meander loops, along the channel margins (natural levees), and at the mouths of distributary channels. Organic rich silts and clays accumulate on the flood plains and in the swamps and marshes. Clays containing marine organisms are deposited in the interdistributary bays and marine shelf. A vertical cross section through a distributary mouth bar is shown at the bottom left of Figure 2.3.

Coarse, unsorted conglomerates and mixed rock types containing much clay, silt, sand, and pebbles are common in alluvial fans. The conglomerate pebbles and rock fragments directly reflect the materials found on the immediately adjacent eroding upland (Figure 2.4). Coarse materials are deposited nearer the head of the fan, and finer sediment more towards the base (Figure 2.5). Near the base sorting improves so that the sediments more nearly resemble typical stream deposits.

Exercise

Three demonstrations will be set up in the laboratory to illustrate map views and cross sections of each of the three major depositional complexes outlined above. Also topographic maps will show examples of these major groups of detrital environments. Examples of all of the major types of detrital rocks will be keyed into their places of origin in the demonstrations. Each student will be given a set of unidentified detrital rocks. Using a binocular microscope or hand lens, the following should be determined for each unknown:

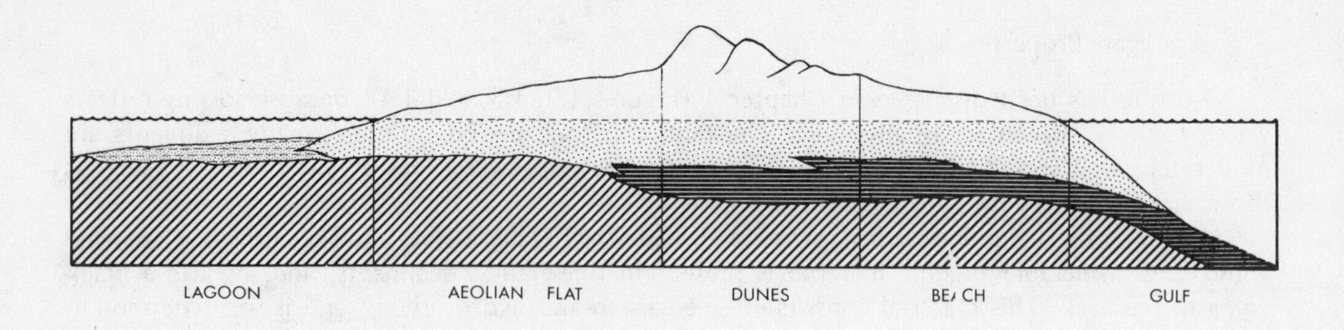

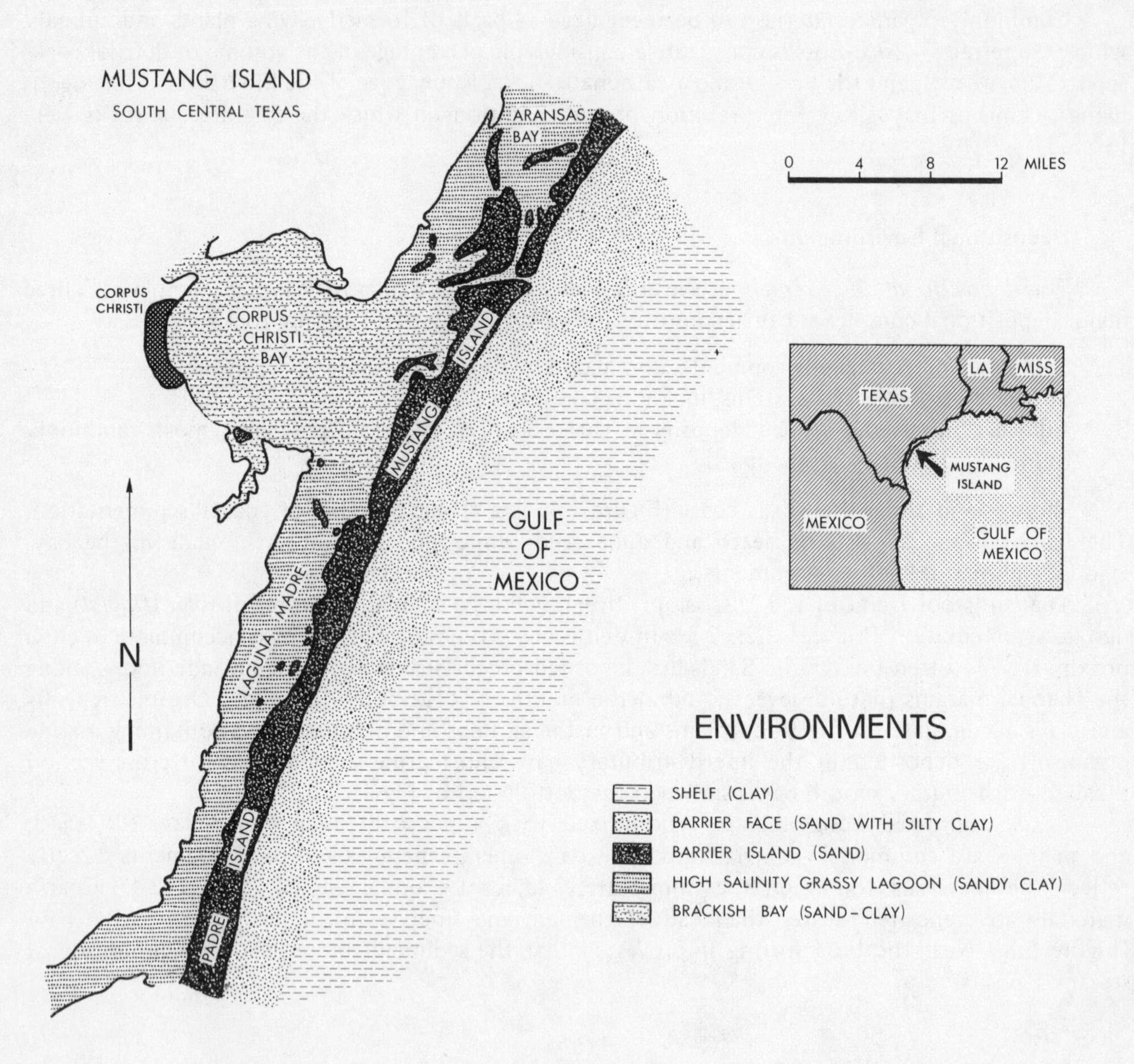

Figure 2.1 Depositional environments associated with a barrier coastline

Figure 2.2 The Birdsfoot Delta of the Mississippi River

Battledore Reef
Sand
Hog Islands
Breton Island
Little Coquille Bay
Grand Coquille Bay
Bird Island
Taylor Pass
Grand Bay
Carencro Bay
Ft Jackson
Boothville
Alexis Bay
Hospital Bay
Yellow Cotton Bay
Spanish Pass
California
Oil
Red Pass
GRAND PASS
MISSISSIPPI RIVER
MAIN PASS
THE MISSISSIPPI DELTA
Fleur Pond
Red Pass
Jaquines Island
Jaquines Pass
Williams Pass
Pelican Bayou
Pass de Wharf
Bayou Tony
Pass Tante Phine
TIGER PASS
Chauee Bay
Oil wells
Joes Bay
Zinzin Bay
Riverside Bay
Bob Taylors Pond
Dixon Bay
Scott Bay
Head of Passes
Bucket Bend
Customhouse Bay
Horse Shoe Pond
RAPHAEL PASS
Bull Bay
Woodyard Pond
Pass a Loutre
North Pass
Middle Ground
North Mud Lumps
Thomasin Lumps
Jackass Bay
BLIND BAY
SOUTHEAST PASS
Cheniere Pass
Redfish Bay
Oil Field
GARDEN ISLAND BAY
SOUTH PASS
Whale Bay
Cockler Bay
EAST BAY
WEST BAY
SOUTHWEST PASS
South Pass Light
South Pass
Oil wells
soft
sand
shells
hard
mud
sticky
36
60
120

a. dominant grain roundness
b. dominant grain size
c. sorting
d. organic content

Determine to which major environmental complex each specimen *most likely* belongs. Make sketches of the map views and cross sections similar to those in the demonstrations, and indicate the probable position(s) of origin of each unknown sedimentary rock.

Test Questions

1. Which detrital rock types are highly diagnostic of particular depositional environments, and which could have been deposited in any of several different environmental settings?
2. What types of fossils (cite specific types) would be particularly useful in determining the depositional environment of some of the less diagnostic rock types?
3. Why are alluvial fan deposits less well sorted than river and stream deposits?
4. In which depositional environments will sorting and roundness be best developed?
5. In which depositional complex might it be possible to find grains of unaltered ferromagnesian minerals and plagioclase?
6. If the rocks of the source area above an alluvial fan have a horizontal stratification of mineral assemblages, in what order will this mineralogical sequence be observed in the adjacent alluvial fan?

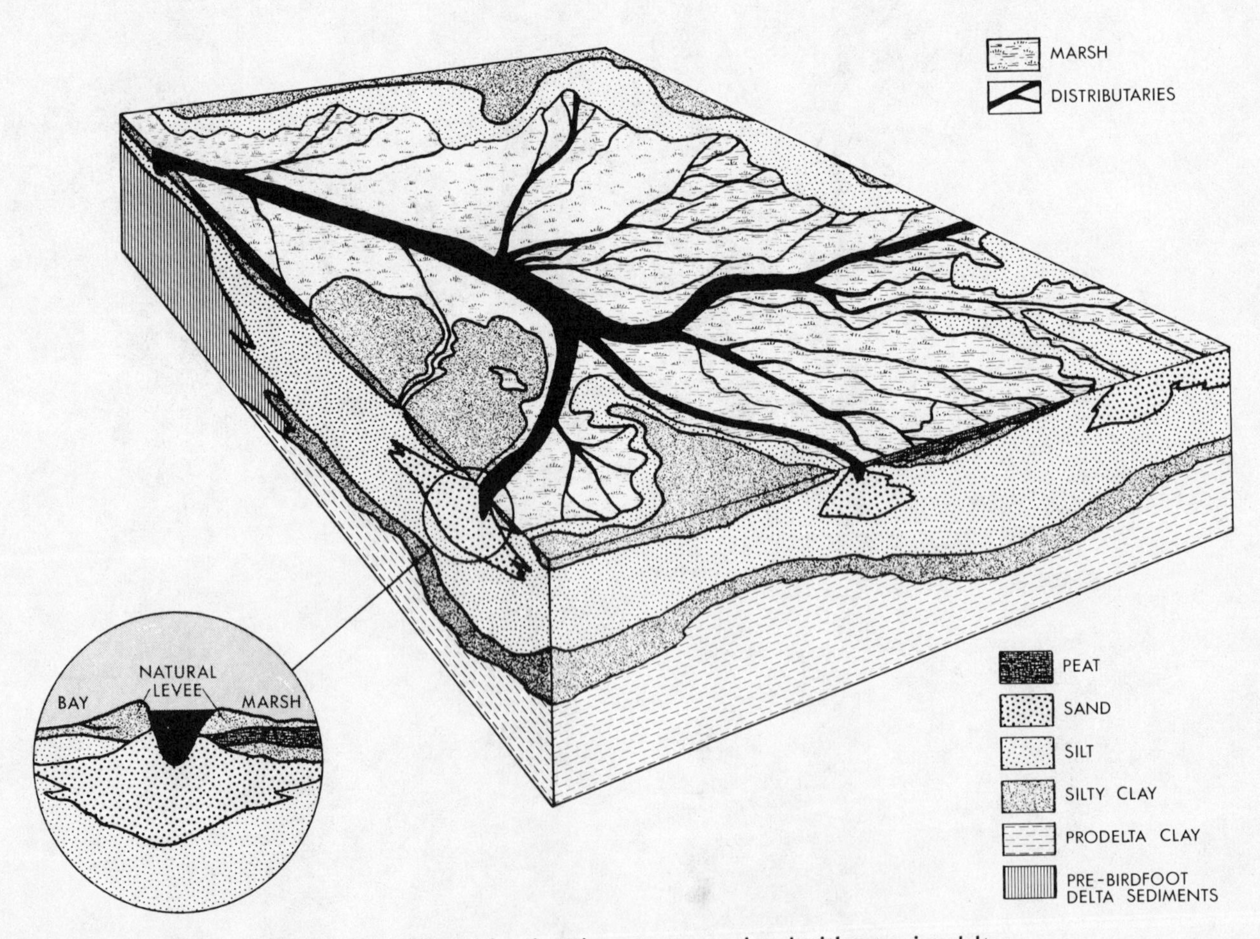

Figure 2.3 Depositional environments associated with a marine delta

Figure 2.4 Stereo photograph illustrating geometry of alluvial fans, and their relationship to their source area.

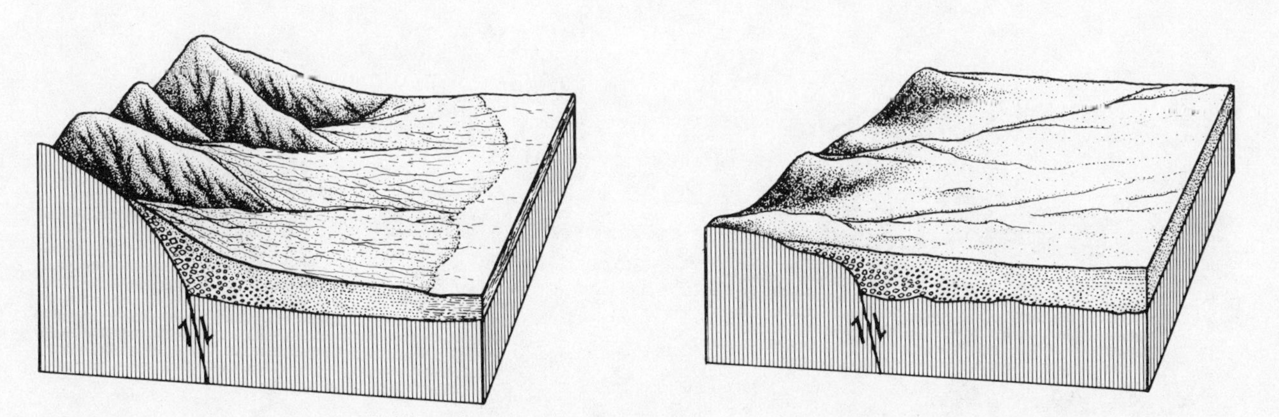

Figure 2.5 Block diagrams showing alluvial fans at successive stages of sedimentation

Background Reading

Bull, W. B., 1972, Recognition of alluvial fan deposits in the stratigraphic record, *in* Rigby, J. K., and Hamblin, W. K., Recognition of Ancient Sedimentary Environments, Soc. Econ. Paleontologists and Mineralogists, Spec. Pub. 16, p. 63-83.

Krauskopf, K. B., 1967, Introduction to Geochemistry, McGraw-Hill Book Co., New York, p. 98-121.

Selley, R. C., 1970, Ancient Sedimentary Environments, Cornell Univ. Press, Ithaca, N.Y., p. 22-51, 74-116.

Chapter Three

Paleontology
Identification of Fossils

Objectives

1. To become acquainted with the differences in skeletal structure among the major groups of organisms preserved as fossils.
2. To view those structural differences in the overall evolutionary framework of the organic world.
3. To understand the basic life habits and habitats of the organisms commonly found as fossils.

Important Terms

Fossils—The preserved remains, usually skeletal fragments, of once living organisms found as the constituent grains of sedimentary rocks.

Life habits—The ways in which an organism goes about its daily activities, primarily those of locomotion and food gathering.

Habitat—The place (either surface or volume) that a particular organism occupies during life.

Discussion

It is known that modern sites of sediment accumulation are characterized by distinctive types of plants, animals, and microorganisms. Those that have preservable tissues will be incorporated into the sediment. Thus ancient environments should be characterized by distinctive fossil assemblages. Because different types of organisms have different tolerances and preferences for environmental conditions, a more embracive idea of an ancient environment can be obtained by identifying the fossils in a rock suite as well as the sediment types. When organisms from different environments are found mixed together, one can infer disruption of sedimentation by mechanisms such as intense storms or hurricanes. By examining changes in form and function from one group of fossils to another in a nearby habitat, one can infer organic evolution.

Because of simple physiological restraints, certain types of organisms have limited tolerance to particular environmental factors, and thus their skeletons are highly indicative of specific habitat conditions. The assemblage of fossils in a rock reflects both the nature of the bottom environment (in subaqueous habitats) and that of the water column overlying the depositional site.

The Evolutionary Framework

Organic life has evolved from simple forms to greater and greater complexity in terms of structure, physiology, growth and development, and behavior. Consequently, organic evolution provides a natural basis for the hierarchical classification of organisms. Evolution can be represented graphically as a "tree of life" (Figure 3.1). The largest or primary branches are the *kingdoms*; the secondary branches, *phyla* (singular, *phylum*); the tertiary branches, *classes*; and so on, down to *orders, families, genera* (singular, *genus*), and the smallest twigs, the *species* (singular and plural, *species*). Consult Stokes (1973, pp. 111-116) for greater detail concerning the biological nature of these groupings.

I. Kingdom MONERA (*procaryotes*—single-celled organisms lacking a nucleus)
    A. BACTERIA (may form stromatolites; see Chapter 9)
    B. CYANOPHYTA (blue-green algae; form stromatolites; see Chapter 9)
II. *Eucaryotes*—Single or multicellular organisms having a nucleus
    A. Kingdom PROTISTA (single-celled eucaryotes)
        1. ALGAE (several types, many found as fossils; see Chapter 6)
        2. Phylum PROTOZOA (several classes found as fossils)
    B. Multicellular Eucaryotes
        1. Kingdom FUNGI (plants lacking chlorophyll)
        2. Kingdom PLANTAE (nonmobile multicellular eucaryotes that photosynthesize; fossil record abundant; several divisions or phyla recognized)
        3. Kingdom ANIMALIA (multicellular eucaryotes which utilize other organisms as food)
            a. Phylum PORIFERA (sponges; lack differentiated organs)
            b. Phylum CNIDARIA (corals, jellyfish, anemones; primitive organ systems)
            c. Coelomates (well-developed body cavity)
                (1) Protostomes (embryonic mouth becomes the mouth of the mature animal)
                    (a) Phylum MOLLUSCA (snails, clams, oysters, scallops, chitons, squids, octopi, etc.)
                    (b) Phylum ARTHROPODA (insects, shrimp, lobsters, crabs, spiders, ticks, etc.)
                (2) Protostome—Deuterostome Intermediates
                    (a) Phylum ECTOPROCTA (bryozoans, "moss" animals)
                    (b) Phylum BRACHIOPODA (lampshells)
                (3) Deuterostomes (embryonic mouth becomes the anus of the mature animal)
                    (a) Phylum ECHINODERMATA (starfish, sea urchins, sea lilies, etc.)
                    (b) Phylum CHORDATA (hemichordates, vertebrates)

Fossil Marine Invertebrates

The skeletons of marine invertebrates are very commonly found as fossils, and constitute the bulk of the sedimentary particles making up ancient limestones. The most common animals whose skeletons are preserved as macrofossils are:

Phylum PORIFERA
    Class STROMATOPOROIDEA (stromatoporoid sponges; Figure 3.2)

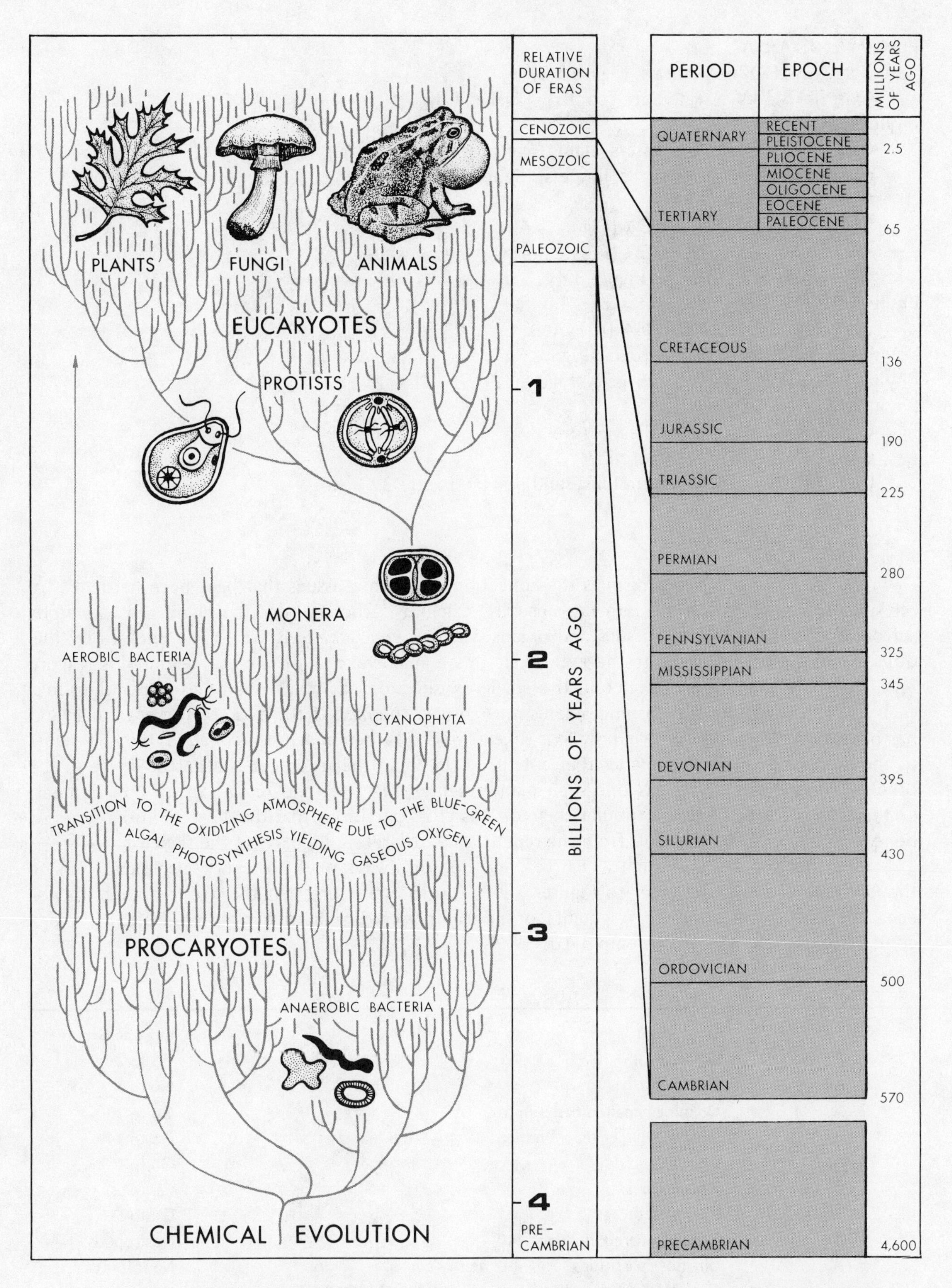

Figure 3.1 Fundamental subdivisions of the organic world and their development in geologic time

Phylum CNIDARIA
- Class ANTHOZOA (corals; Figure 3.2)
- Class TABULATA (tabulate "corals"; Figure 3.2)

Phylum MOLLUSCA
- Class PELECYPODA (oysters, clams, scallops; Figure 3.3)
- Class GASTROPODA (snails, limpets; Figure 3.3)
- Class CEPHALOPODA
  - Order NAUTILOIDEA (Figure 3.4)
  - Order AMMONOIDEA (Figure 3.4)
  - Order SEPIOIDEA (Figure 3.4)

Phylum ARTHROPODA
- Class TRILOBITOMORPHA (trilobites, Figure 3.5)

Phylum ECTOPROCTA (bryozoans; Figure 3.6)

Phylum BRACHIOPODA (lampshells; Figure 3.6)

Phylum ECHINODERMATA
- Class CRINOIDEA (sea lilies; Figure 3.7)
- Class BLASTOIDEA (Figure 3.7)
- Class ECHINOIDEA (sea urchins, sand dollars; Figure 3.5)

Fossil Identification Key

The *skeleton* of an organism is the collection of organic tissues that have been hardened by physiologic impregnation of minerals, usually calcium carbonate or phosphate. *Exoskeletons* surround most of the other tissues and organs, whereas *endoskeletons* are totally encased by the other tissues. Most invertebrates have exoskeletons; a major exception are the belemnoids (Figure 3.4), that have only an endoskeleton. If a single exoskeleton has numerous tubes or openings that each housed a separate functioning organism, then it is *colonial*. If the exoskeleton is tubular and has been turned around an imaginary axis (like a twisted garden hose), it is *coiled*. *Symmetry* refers to the number of imaginary planes that cut the skeleton so that each cut is a mirror image of the other. *Bilateral* symmetry has one such plane; *pentameral*, five; and *radial*, an infinite number. *Chambers* are formed when a tubular skeleton has cross partitions that divide it up into a series of hollow sections. *Segments* result from the repetition of discrete solid parts of the skeleton.

Identification keys are highly abstract, and do not always work for every specimen. In using the key below, constantly refer to Figures 3.2 to 3.6 to make sure that the wrong branch has not been taken. Identification of fragments or poorly preserved material is always difficult, and a combination of tactics must be employed.

Key Category	Description	Organic Group
1. A.	Colonial	Go to 2
B.	Noncolonial	Go to 3
2. A.	Openings smaller than 1 mm	Go to 13
B.	Openings larger than 1 mm	Go to 14
3. A.	Coiled	Go to 4
B.	Noncoiled	Go to 6
4. A.	Chambered	Go to 5
B.	Nonchambered, hollow	Gastropods
5. A.	Chamber partitions straight or gently curved	Nautiloids
B.	Chamber partitions highly curved with "crinkly" edges	Ammonoids

Key Category	Description	Organic Group
6. A.	Bilateral symmetry	Go to 7
B.	Pentameral symmetry	Go to 9
C.	Radial symmetry	Go to 11
7. A.	Two paired shells (valves)	Go to 8
B.	Segmented, having three lobes	Trilobites
8. A.	Paired shells exact mirror images	Pelecypods
B.	Shell halves not mirror images; each valve has bilateral symmetry	Brachiopods
9. A.	Globular or vaselike, covered by polygonal plates; stem remnants or attachment marks at base	Go to 10
B.	Globular or disklike, not connected to a stem	Echinoids
10. A.	Five radiating elaborate grooves	Blastoids
B.	No elaborate grooves	Crinoids
11. A.	Conical or tapering shell	Go to 12
B.	Cylindrical, composed of disklike segments	Crinoid stems
12. A.	Horizontal partitions forming chambers	Go to 5
B.	Vertical partitions radially projecting inward	Solitary corals
C.	Shell solid, cigarlike, having radial orientation of fibrous crystals	Belemnites
13. A.	Massive, with thin laminations; each thin layer having numerous boxlike chambers	Stromatoporoids
B.	Not thinly laminated	Bryozoans
14. A.	Vertical partitions radially projecting inward within each opening	Colonial corals
B.	Horizontal partitions forming chambers	Tabulate "corals"

Life Habits and Habitats

The major environments occupied by organisms are shown in Figure 3.7. Organisms which float freely in the water (usually near the water's surface with the air) are termed *plankton* (plants are *phytoplankton*; animals, *zooplankton*). Those animals that swim, or control their own sense of movement in the water column, are termed *nekton* (adjective, *nektonic*). Animals and plants intimately associated with the substratum, either unconsolidated sediment or a hard rocky seafloor, are the *benthos* (adjective, *benthonic*). Those which are attached and cannot move about freely are *sessile*, whereas those that crawl about over the surface are *vagrant*. Benthonic animals above the sediment-water interface are the *epifauna*, whereas animals that live within the sediment (burrowers) are the *infauna*. The interrelationships of these habitats are illustrated in Figure 3.8. The organisms that occupy each major type of habitat are shown in Figures 3.9 to 3.17.

Another major life habit is the means of obtaining food. Feeding mechanisms determine the role played by the organism within its community. Plants manufacture their own food through photosynthesis, and are termed *autotrophs*. Only autotrophs produce food; all other organisms present are consumers. If animals depend mainly upon plankton and suspended food, they are *filterers*. Organisms which mine organic debris that has accumulated within the substrate are *deposit feeders*. Organisms that feed upon plants are *herbivores* and those that feed upon animals are *carnivores*. Feeders that search out dead organic matter lying on top of the sediment are *scavengers*. The interrelationships of these feeding types are shown in Figure 3.8. The table on page 30 indicates the life habits of marine organisms commonly found as macrofossils (microscopic fossils and vertebrates have been omitted).

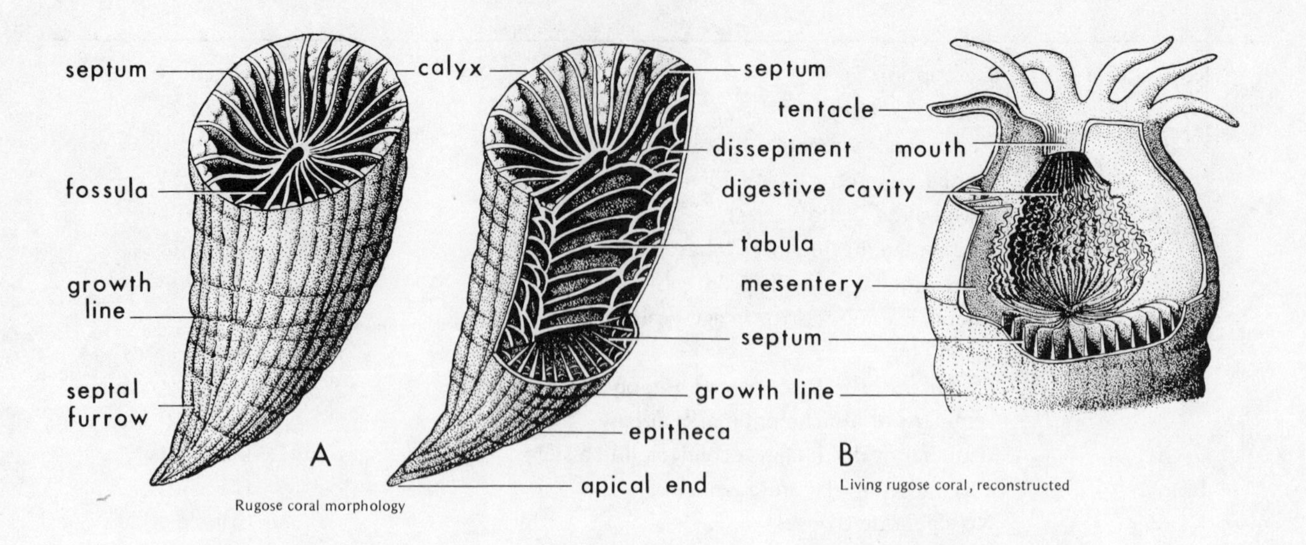

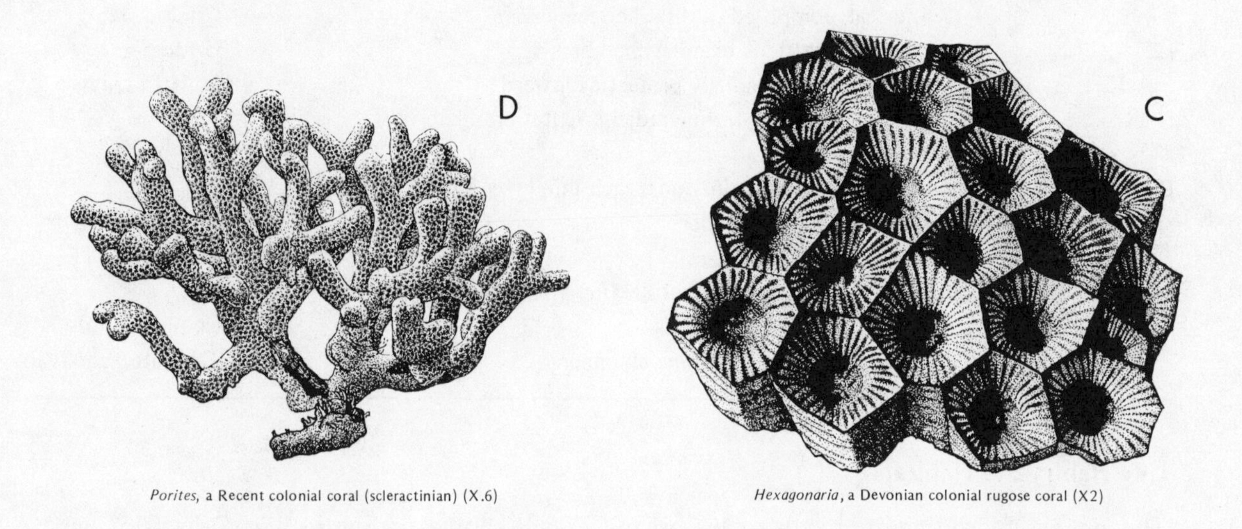

Porites, a Recent colonial coral (scleractinian) (X.6)

Hexagonaria, a Devonian colonial rugose coral (X2)

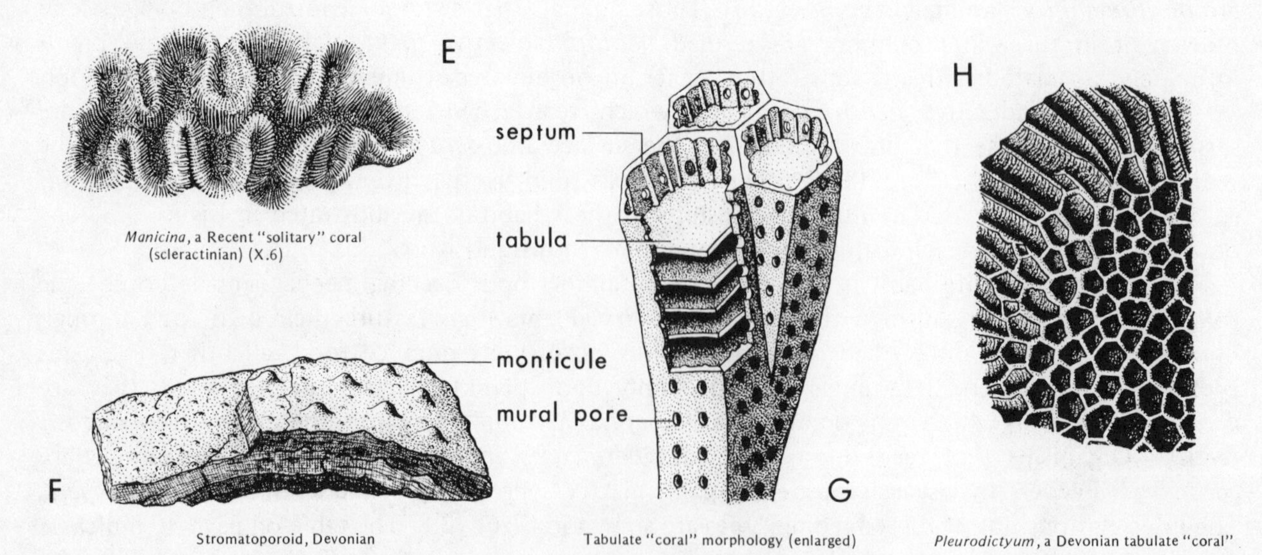

Manicina, a Recent "solitary" coral (scleractinian) (X.6)

Stromatoporoid, Devonian

Tabulate "coral" morphology (enlarged)

Pleurodictyum, a Devonian tabulate "coral"

Figure 3.2 Corals and stromatoporoids

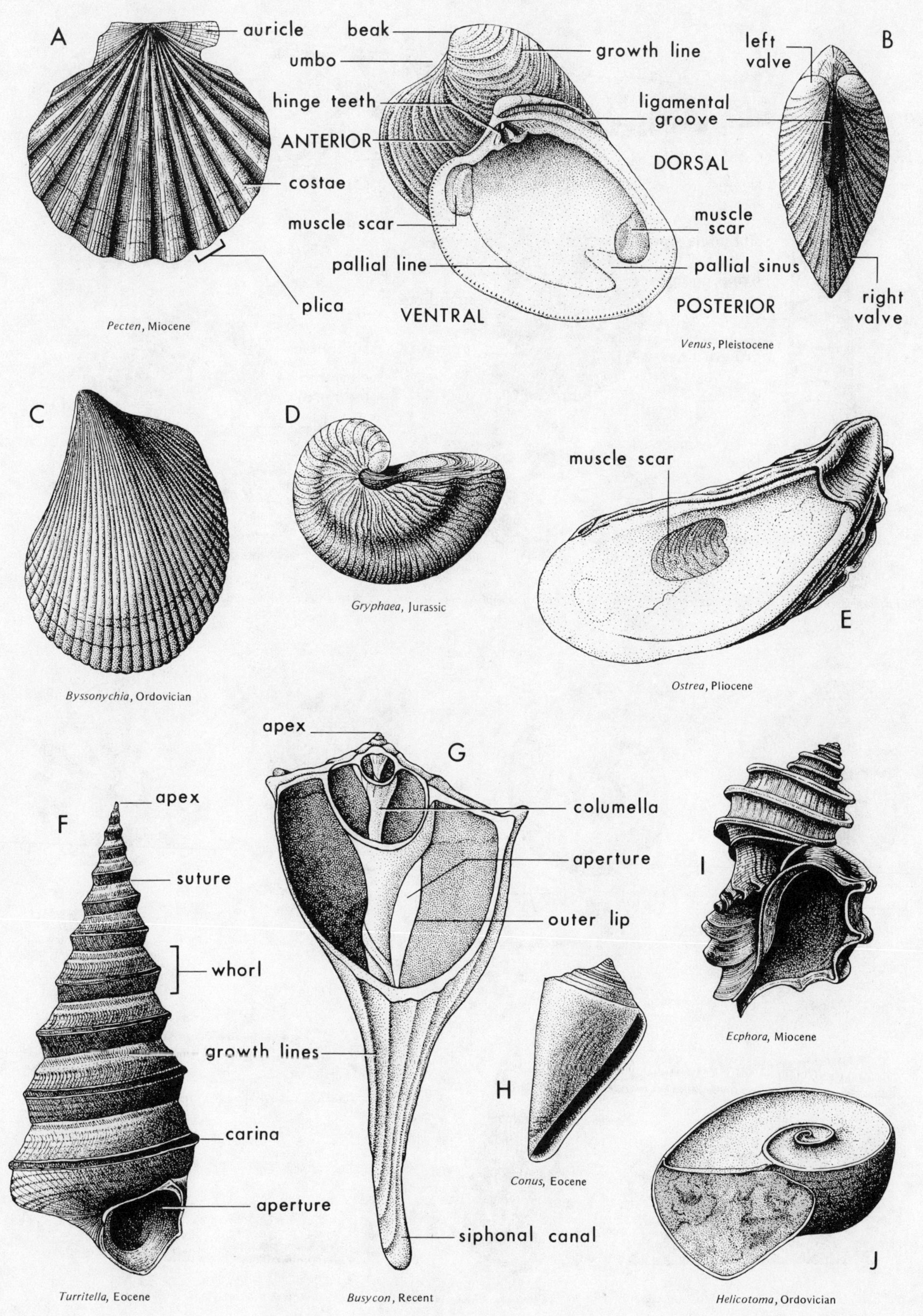

Figure 3.3 Pelecypods (A-E) and gastropods (F-J)

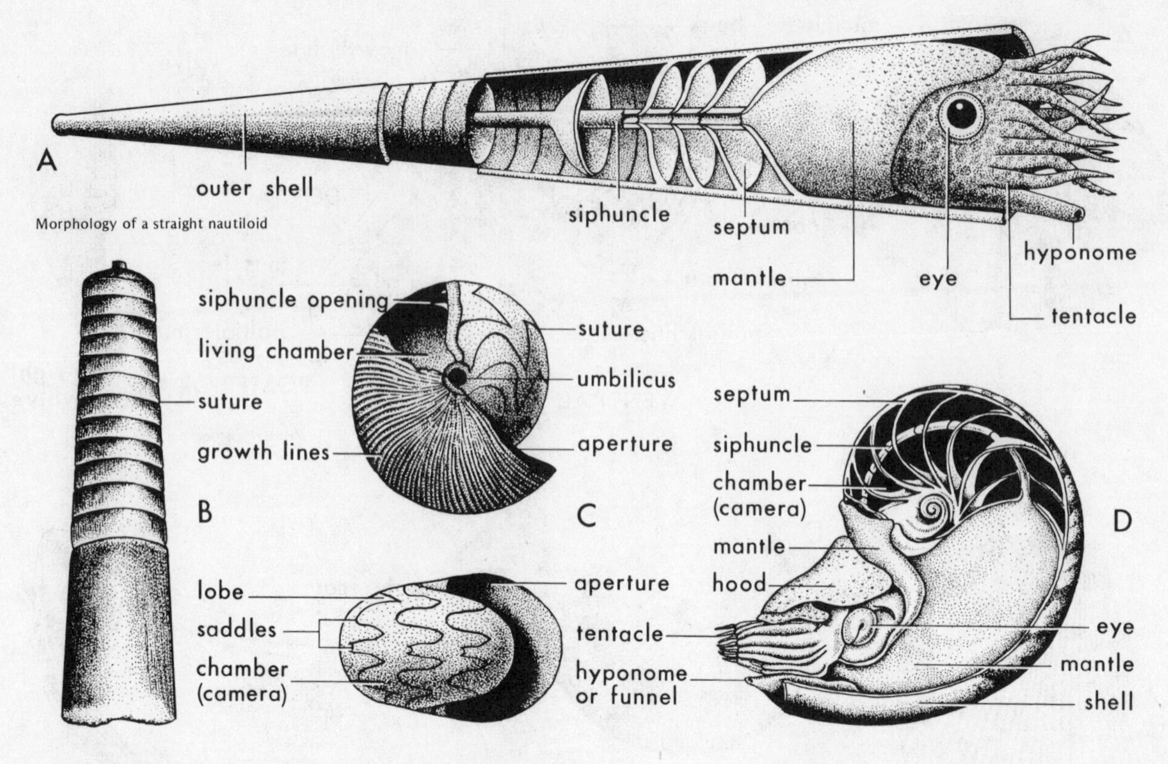

Morphology of a straight nautiloid

Michelinoceras, Ordovician, a straight nautiloid

Morphology of a Mississippian goniatite

Morphology of *Nautilus*, a Recent nautiloid

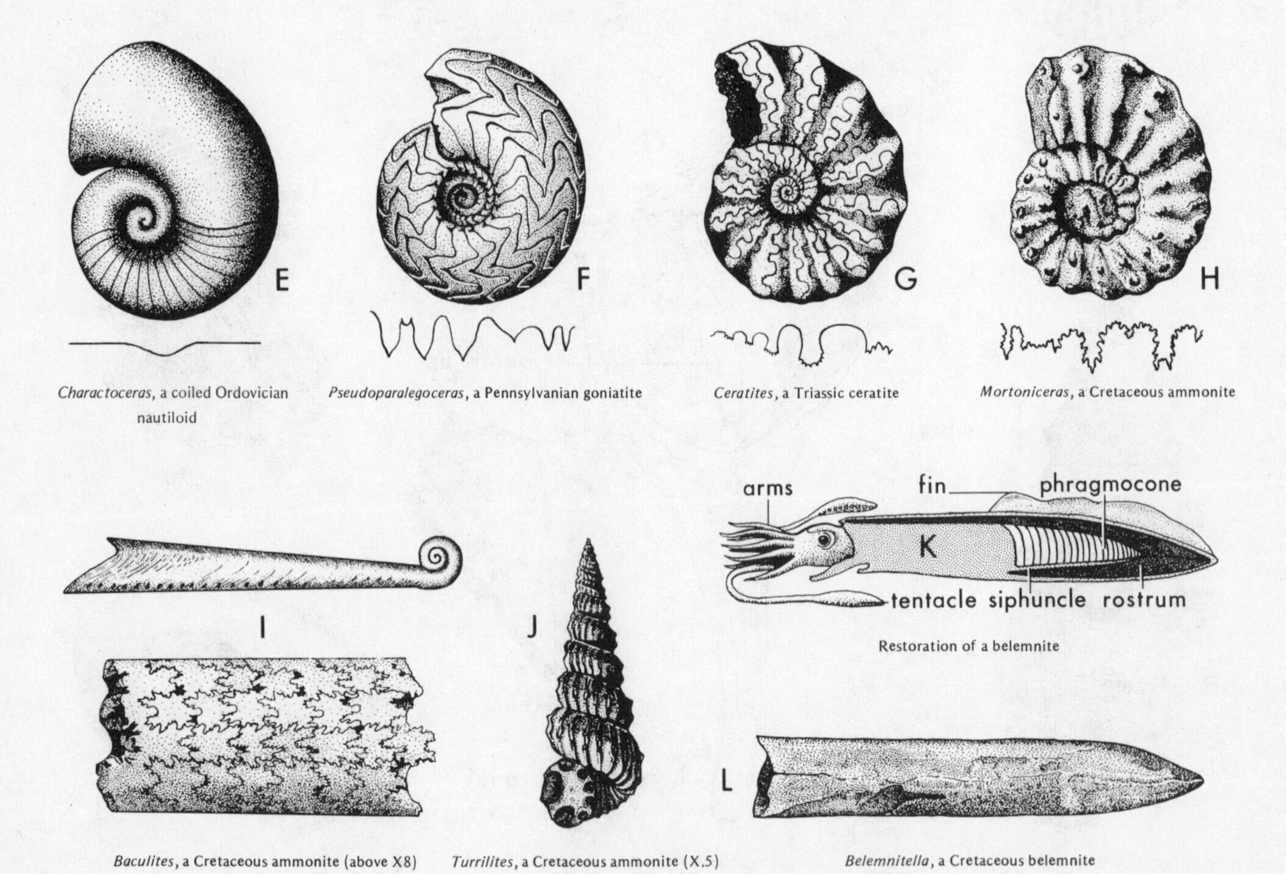

Charactoceras, a coiled Ordovician nautiloid

Pseudoparalegoceras, a Pennsylvanian goniatite

Ceratites, a Triassic ceratite

Mortoniceras, a Cretaceous ammonite

Restoration of a belemnite

Baculites, a Cretaceous ammonite (above X8)

Turrilites, a Cretaceous ammonite (X.5)

Belemnitella, a Cretaceous belemnite

Figure 3.4 Cephalopods

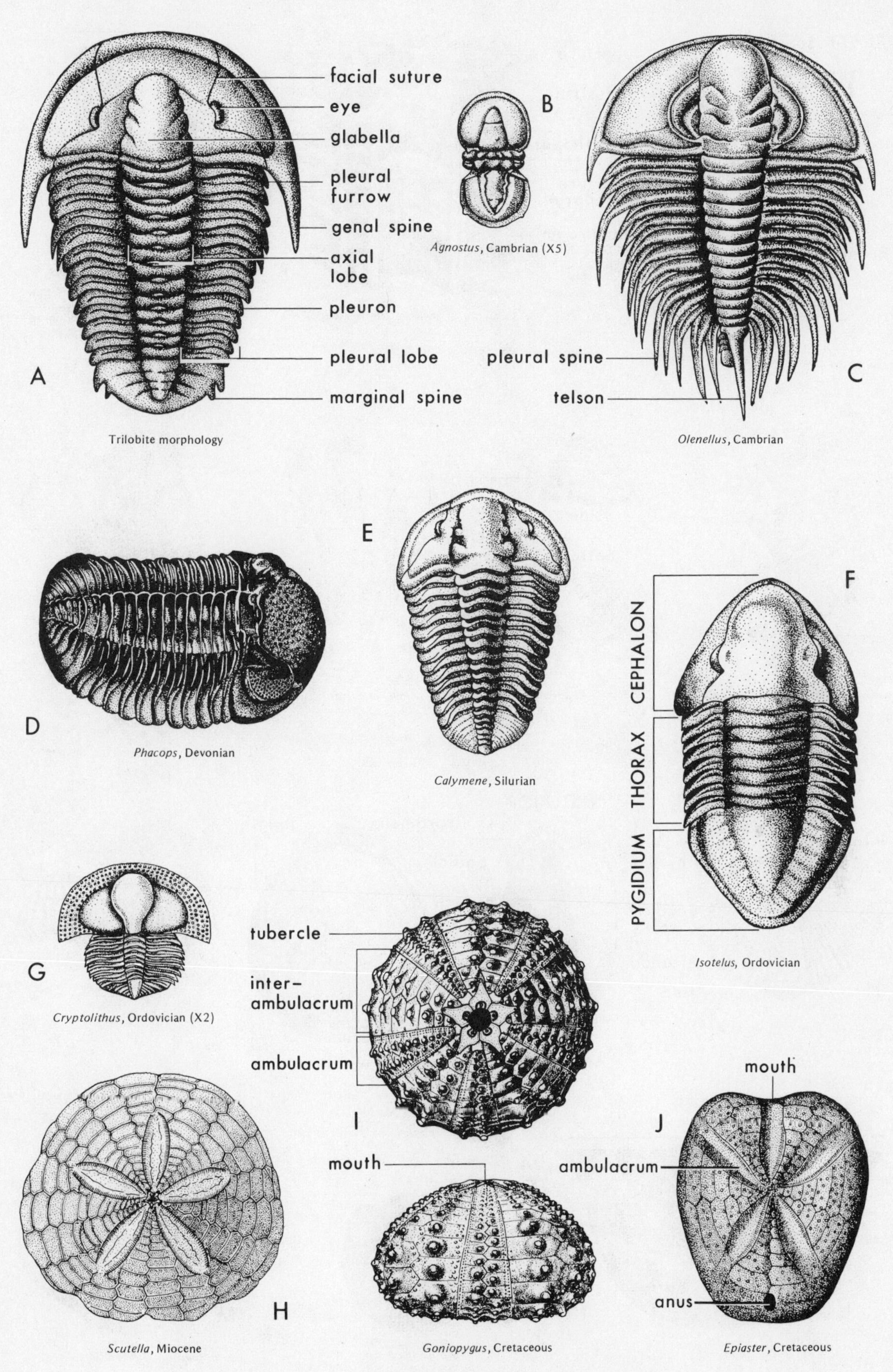

Figure 3.5 Trilobites (A-G) and echinoids (H-J)

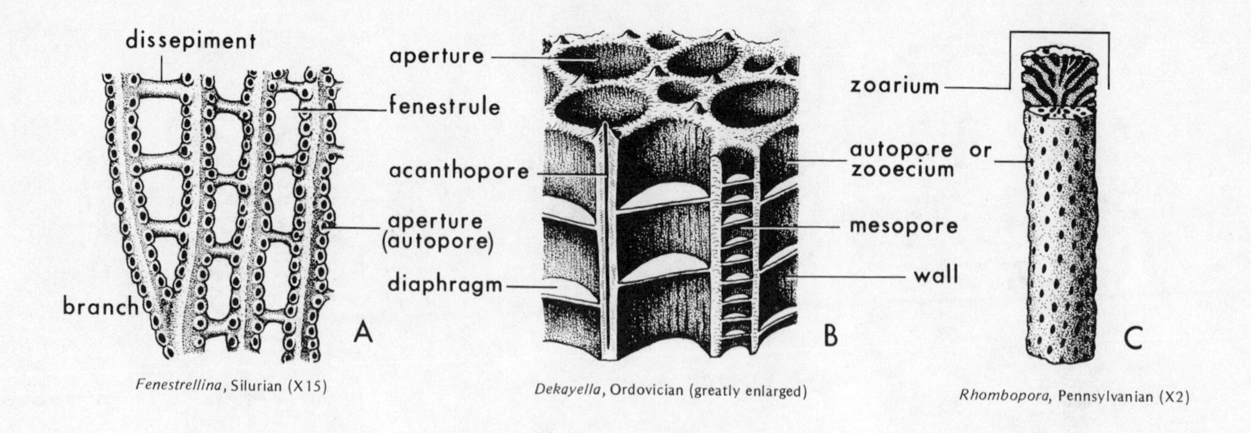

Fenestrellina, Silurian (X15)

Dekayella, Ordovician (greatly enlarged)

Rhombopora, Pennsylvanian (X2)

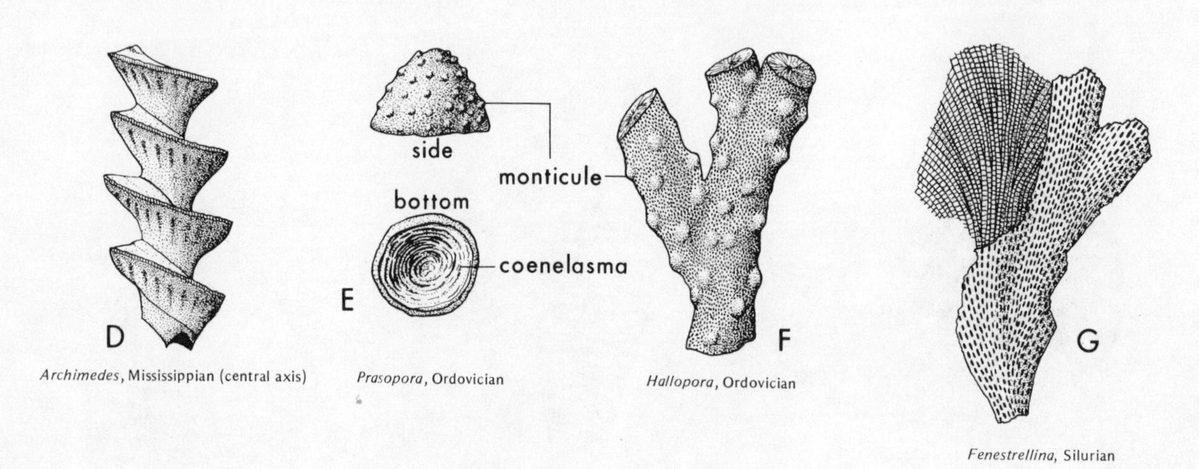

Archimedes, Mississippian (central axis)

Prasopora, Ordovician

Hallopora, Ordovician

Fenestrellina, Silurian

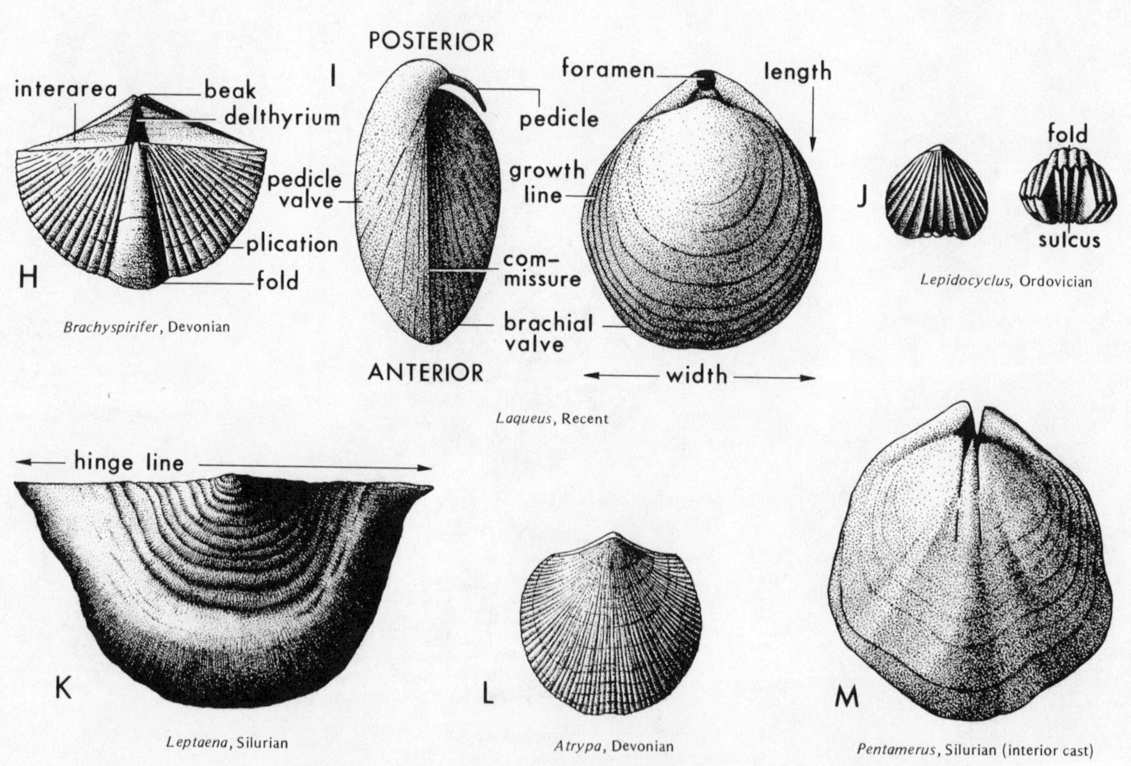

Brachyspirifer, Devonian

Laqueus, Recent

Lepidocyclus, Ordovician

Leptaena, Silurian

Atrypa, Devonian

Pentamerus, Silurian (interior cast)

Figure 3.6. Bryozoans (A-G) and brachiopods (H-M)

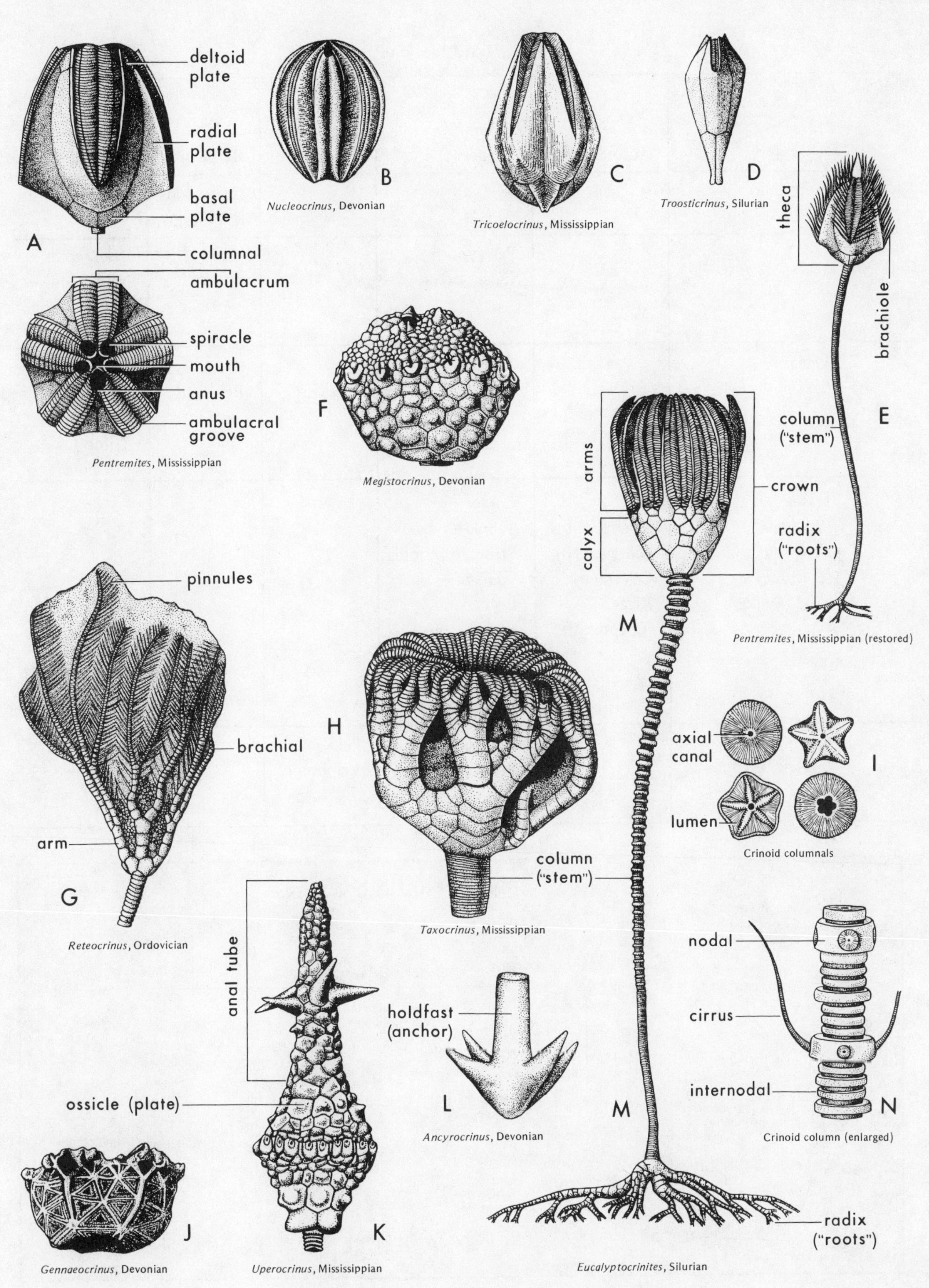

Figure 3.7 Blastoids (A-E) and crinoids (F-N)

TABLE 1

Habitat \ Feeding Type	Autotrophs	Filter Feeders	Sediment Feeders	Herbivores	Carnivores / Scavengers
Planktonic				Gastropods (Pteropods)	
Nektonic		Pelecypods (Scallops)			Belemnoids Nautiloids Ammonoids Trilobites
Vagrant Benthonic		Trilobites		Gastropods Echinoids	Gastropods Nautiloids Ammonoids Trilobites Echinoids
Sessile Benthonic Epifauna (& Epiphyta)	Algal Stromatolites (Chapter 10) Calcareous Algae (Chapter 7)	Corals Tabulates Stromatoporoids Sponges Bryozoans Brachiopods Pelecypods Crinoids Blastoids			Corals
Benthonic Infauna		Pelecypods Brachiopods	Pelecypods Echinoids		Gastropods Trilobites

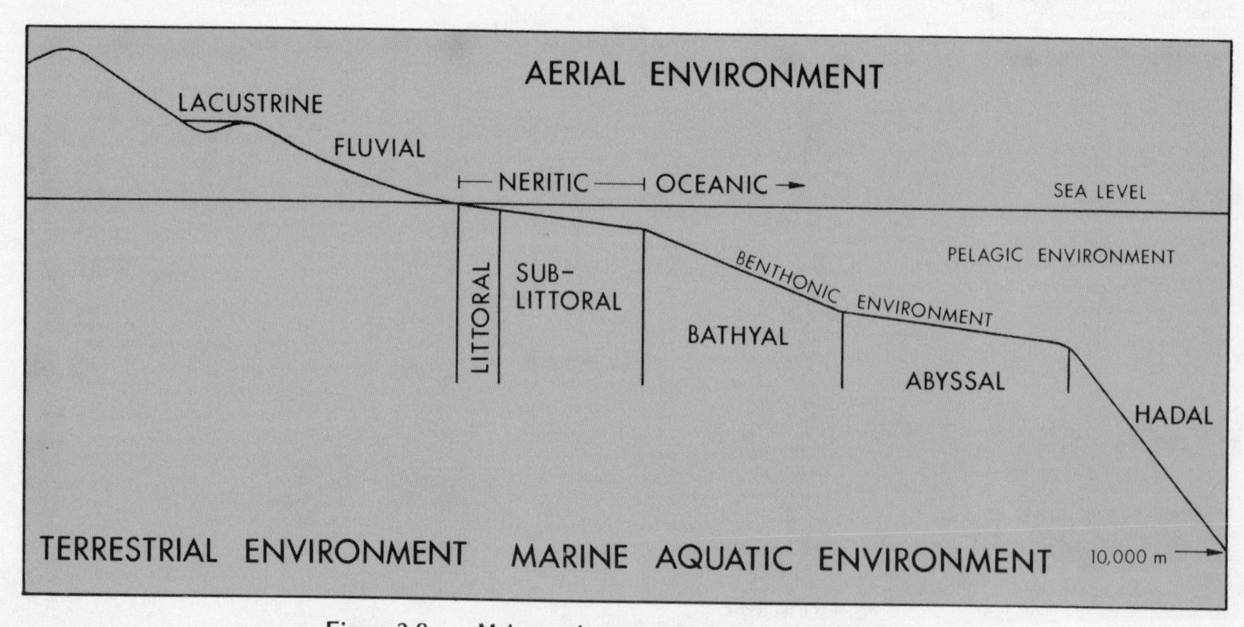

Figure 3.8 Major environments occupied by organisms

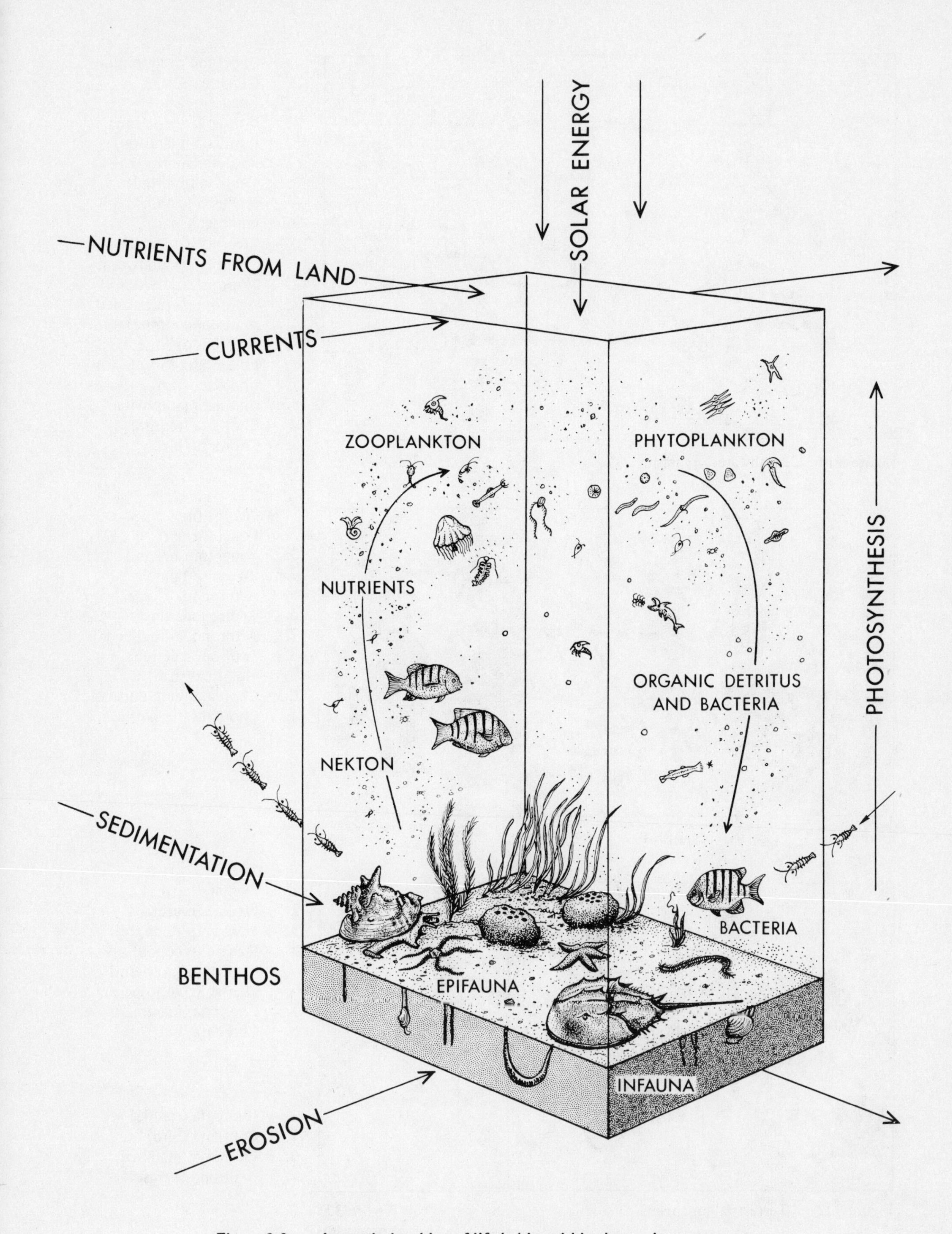

Figure 3.9 Interrelationships of life habits within the marine ecosystem

1. Algae and cyanophyta
2. Bacteria
3. Fungi (mold)
4. Plantae (angiosperm)
5. Protozoa (plankton)
6. Porifera (sponge)
7. Cnidaria (hydroid)
8. Mollusca (clam)
9. Mollusca (snail)
10. "Worms" (annelid)
11. "Worms" (roundworm)
12. "Worms" (flatworm)
13. Arthropoda (crayfish)
14. Arthropoda (mite)
15. Arthropoda (insect)
16. Ectoprocta (bryozoan)
17. Chordata (fish)
18. Chordata (amphibian)
19. Chordata (reptile)
20. Chordata (bird)

Figure 3.10 Freshwater organisms

1. Plantae (moss)
2. Fungi (lichen)
3. Fungi (mushroom)
4. "Worms" (annelid)
5. Arthropoda (spider)
6. Arthropoda (insect)
7. Arthropoda (millipede)
8. Arthropoda (crab)
9. Mollusca (snail)
10. Chordata (amphibian)
11. Chordata (reptile)

Figure 3.11 Terrestrial organisms

1. Plantae (fern)
2. Plantae (horsetail)
3. Plantae (psilophyte)
4. Plantae (lycopod)
5. Plantae (angiosperm)
6. Plantae (gymnosperm)
7. Chordata (mammal)
8. Chordata (bird)

Figure 3.12 Terrestrial organisms

1. Chordata (reptile)
2. Chordata (bird)
3. Chordata (mammal)
4. Arthropoda (insect)

Figure 3.13
Aerial organisms

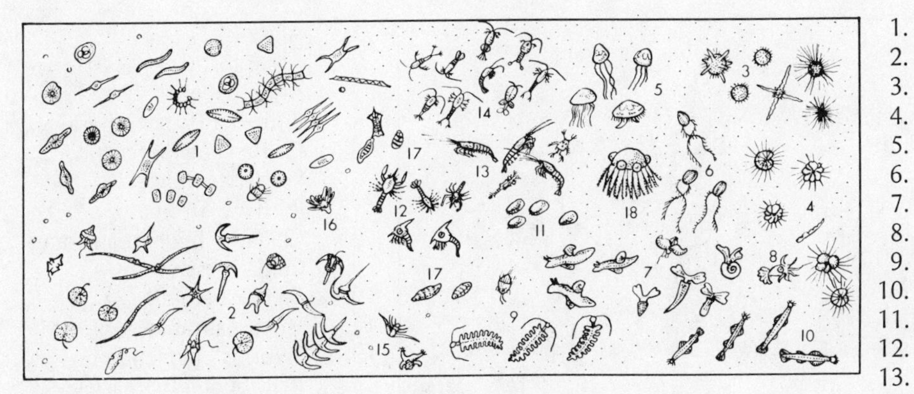

Figure 3.14 Planktonic marine organisms

1. Algae (diatoms)
2. Algae (flagellates)
3. Protozoa (radiolarians)
4. Protozoa (foraminifera)
5. Cnidaria (medusae)
6. Cnidaria (comb-jellyfish)
7. Mollusca (snails)
8. Mollusca (nautiloid)
9. "Worms" (annelid)
10. "Worms" (arrowworm)
11. Arthropoda (ostracods)
12. Arthropoda (crab larvae)
13. Arthropoda (krill)
14. Arthropoda (copepods)
15. Echinodermata (starfish larvae)
16. Echinodermata (sea urchin larvae)
17. Chordata (tunicates)
18. Chordata (graptolite)

1. Cnidaria (jellyfish)
2. Mollusca (nautiloid)
3. Mollusca (squids)
4. Arthropoda (shrimp)
5. Arthropoda (eurypterid)
6. Chordata (fish)
7. Chordata (reptile)
8. Chordata (mammal)

Figure 3.15 Nektonic marine organisms

1. Cnidaria (anemones)
2. Mollusca (snails)
3. Mollusca (chiton)
4. Mollusca (scallop)
5. Mollusca (octopus)
6. Mollusca (nautiloid)
7. Mollusca (ammonite)
8. "Worms" (ribbon worm)
9. "Worms" (annelid)
10. Arthropoda (trilobite)
11. Arthropoda (crab)
12. Arthropoda (eurypterid)
13. Echinodermata (starfish)
14. Echinodermata (sea cucumber)
15. Echinodermata (sea urchin)
16. Echinodermata (brittle star)
17. Chordata (fish)

Figure 3.16 Marine vagrant benthonic organisms

1. Algae (red)
2. Algae (brown)
3. Algae (green)
4. Protozoa (foraminifera)
5. Plantae (angiosperm)
6. Porifera (sponge)
7. Porifera (stromatoporoid)
8. Cnidaria (hydrozoan coral)
9. Cnidaria (sea whip)
10. Cnidaria (coral)
11. Mollusca (mussels)
12. "Worms" (annelid)
13. Arthropoda (barnacles)
14. Ectoprocta (bryozoan)
15. Brachiopoda (lampshells)
16. Echinodermata (crinoids)
17. Echinodermata (blastoids)
18. Echinodermata (cystoid)

Figure 3.17 **Marine sessile benthonic organisms (epifauna)**

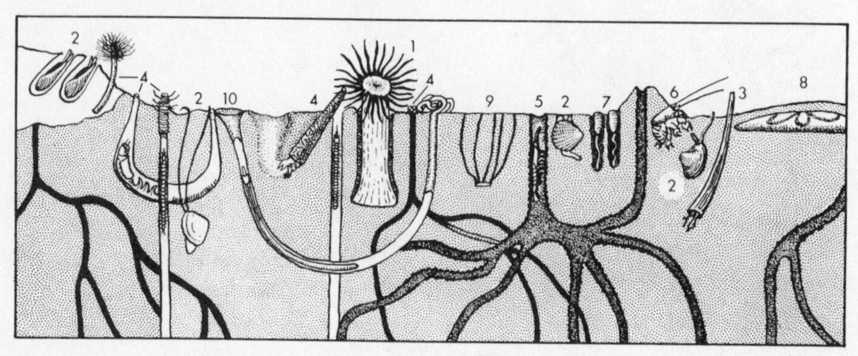

1. Cnidaria (burrowing anemone)
2. Mollusca (clams)
3. Mollusca (tusk shell)
4. "Worms" (annelids)
5. Arthropoda (ghost shrimp)
6. Arthropoda (mole crab)
7. Brachiopoda (inarticulates)
8. Echinodermata (sand dollar)
9. Echinodermata (brittle star)
10. Chordata (acorn worm)

Figure 3.18 **Marine benthonic organisms (infauna)**

Different physical environments may be distinguished from each other by different proportions of feeding types and habitat types in a preserved fossil assemblage. The abundance of autotrophs, filterers, and herbivores in an assemblage reflects how well lighted the water was, because all of these organisms depend directly upon photosynthesis for their food supply. The abundance of scavengers and sediment feeders may reflect the absence of photosynthetic food sources. The abundance of filter feeders reflects the cleanness of the water (lack of suspended sediment that would clog feeding mechanisms). The abundance of sessile benthonic epifauna reflects substrate firmness, because these organisms require relatively hard surfaces of attachment. To summarize, life habits reflect depth of light penetration (water depth), turbidity (rate of sedimentation), and firmness of substrate. Also, food chain relationships and relative productivity can be determined from the ratios of carnivores to herbivores and organisms lower in the food pyramid.

Not all of the organisms in the table above are restricted to one habitat or feeding type. Additional information is needed to characterize the life habits of any given fossil. This is often obtainable from the shell form. Plankton must be spherical or globular, to function best as floaters. Buoys or inner tubes would be good physical analogues. Conversely, nekton should be bilaterally symmetrical and streamlined, as are ships and submarines. Benthonic epifauna need attachment structures, such as foundations, anchors, or roots. Infaunal elements must be shaped like shovels or prods, or have sawlike ridges and grooves for slicing through soft sediment.

Exercise

Several large slabs bearing assemblages of fossils have been set out in the laboratory.

1. Using the identification key, identify all of the fossil types present in each slab. Examine each slab under a low power microscope to find small forms.
2. Determine the life habit of each fossil form, using Figures 3.8 to 3.17, the table above, and the shell form. State your reasons for each assignment based on shell form alone. Two or more life habits may be present among fossils belonging to the same phylum or class.
3. What were the environmental conditions, as closely as can be determined, in which each assemblage lived? Rank your assemblages in order of light penetration, turbidity, substrate firmness, and productivity. Which assemblages lived in shallow, nearshore environments, and which lived in deeper, offshore habitats?

Test Questions

1. What analogies can be drawn between the shells of particular animals you have seen in the laboratory and simple machines or mechanical designs that man has contrived to perform a similar function?
2. Which fossil groups have only one feeding type and habitat? Which have more than one? What difference in shell form did you observe among fossils of the same organic group that had different life habits?
3. What additional aspects of the ancient environment could be determined from analysis of life habits?

Background Reading

Laporte, L. F., 1968, Ancient Environments, Prentice-Hall, Englewood Cliffs, N.J., p. 30-60.
McAlester, A. L., 1968, The History of Life, Prentice-Hall, Englewood Cliffs, N.J., p. 39-62.
Stokes, W. L., 1973, Essentials of Earth History, 3rd ed., Englewood Cliffs, N.J., p. 84-117.

Chapter Four

Biostratigraphy
Geologic Time

Objectives

1. To gain an appreciation of the magnitude of geologic time, and the time relationships of the cardinal events of earth history.
2. To learn one of the methods used by geologists to determine the ages of sedimentary rocks and fossil assemblages.
3. To identify, as precisely as possible, the geologic ages of unknown slabs of rock containing diverse fossil assemblages.

Important Terms

Eon—The largest subdivision of geologic time; the rocks deposited during an eon are termed an *eonothem*. The *Phanerozoic* Eon constitutes the last 600 million years of earth history, and represents the time of the evolution of multicellular plants and animals.

Era—The largest subdivision of an eon; the rocks deposited during an era are termed an *erathem*. The Phanerozoic Eon consists of the *Paleozoic*, *Mesozoic*, and *Cenozoic* Eras (Figure 3.1).

Period—The largest subdivision of an era; the rocks deposited during a period are termed a *system*.

Epoch—The largest subdivision of a period; the rocks deposited during an epoch are termed a *series*.

Correlation—The process of determining the time relationships within an undated body of rock by relating it to a dated sequence, using either radiometric dating or fossil types.

Superposition—The observation that the lower strata in an undisturbed succession of sedimentary rocks are older than those overlying them. Faults and igneous intrusions may cut through sedimentary rocks, and are younger than the rocks they cut but older than rocks that truncate them erosionally.

Isochron—A surface or layer that represents points of simultaneous deposition within a body of sedimentary rocks. Examples are volcanic ash deposits (bentonites), varves in lake deposits, storm-deposited beds, lava flows, or the result of any depositional process that could *simultaneously* affect a large geographic area.

Discussion

The determination of geologic time presents one of the most difficult problems in geology. In some areas excellent isochrons, such as extensive beds of volcanic ash, are present. Unfortunately such features are relatively rare, and usually are of little aid in correlation across large geographic areas, such as across an entire continent, or from one continent to another. When igneous rocks cut across sedimentary rocks, as do dikes, sills, or surface flows, radiometric age dates can determine either an upper or lower age limit for the sediments, but usually cannot date them precisely. Geologically young sediments can be dated directly using the decay of carbon-14 (only back to 40,000 years ago) and thorium-230 (only back to 300,000 years ago). Glauconite, a mineral silicate of potassium, aluminum, and iron, often crystallizes within marine sediments at their time of deposition, and can provide a direct radiometric measurement of the age of the sediment, using the potassium-argon decay method. However, many sediments do not contain glauconite, and the loss of argon with time causes the age dates generally to be too young. Most commonly, however, these direct methods of age determination cannot be carried out, and the geologist must resort to indirect methods, primarily using fossils.

Physical stratigraphy, or the tracing out of beds of uniform lithology across a geographic area, is of very little value in determining time relationships, because a homogeneous rock unit usually represents a single depositional environment that has migrated spatially, so that the same rock unit represents deposition at different times in different geographic areas. In fact, a greater duration of time may be present in the lateral development of a homogeneous rock unit across a large area than in the vertical sequence of superposed rock units at a single locality. The types of stratigraphic simulation models used in Chapters 12 through 17 may be extremely valuable in determining the time relationships within large bodies of sedimentary rock.

Because of the general absence of any fossils and of multicellular organisms in particular, time correlation is exceptionally difficult in the eonothems preceding the Phanerozoic. Only the Phanerozoic Eonothem contains abundant fossils of animal life, and their time relationships are used for dating unknown sequences. Figure 4.1 shows the known time ranges of the fossil types studied in Chapter 3. The use of a single fossil type in a rock, such as gastropods, does not clearly specify any time period. However, the presence of several fossil types together in the same rock unit brackets the time period more closely. For example, a rock containing ammonoids, belemnites, and trilobites could only have been deposited during the interval from the middle Carboniferous to the late Permian. This method of time determination is known as the *overlapping range method*. The precision of this method depends upon the accuracy of the fossil identifications as well as accurate knowledge of their time ranges. The animals listed in Figure 4.1 are identified only to the phylum or class level, and thus their time ranges are very long. The more precisely a fossil is identified, from order and family down to genera and species, the narrower are the time limits that it specifies. Most species have a range less than one epoch; genera, less than a period; families, less than an era. Orders, classes, and phyla have, for the most part, longer ranges in time.

Exercise

Part I

Because of the great magnitude of geologic time, it is difficult to understand the proper position in time of the various key events that are studied in earth history. It is a useful exercise to convert these time values to more commonplace or everyday time units in order to make them more comprehensible. The first part of this exercise is to apportion all of geologic time, 4.5 billion years, to a single year, and to determine the dates of certain key events in earth history within that year. Consider the origin of the earth to be January 1, and the present day to be midnight of December 31. Determine the dates of the following geologic events within this year of years, and for events on December 31, determine their time of day:

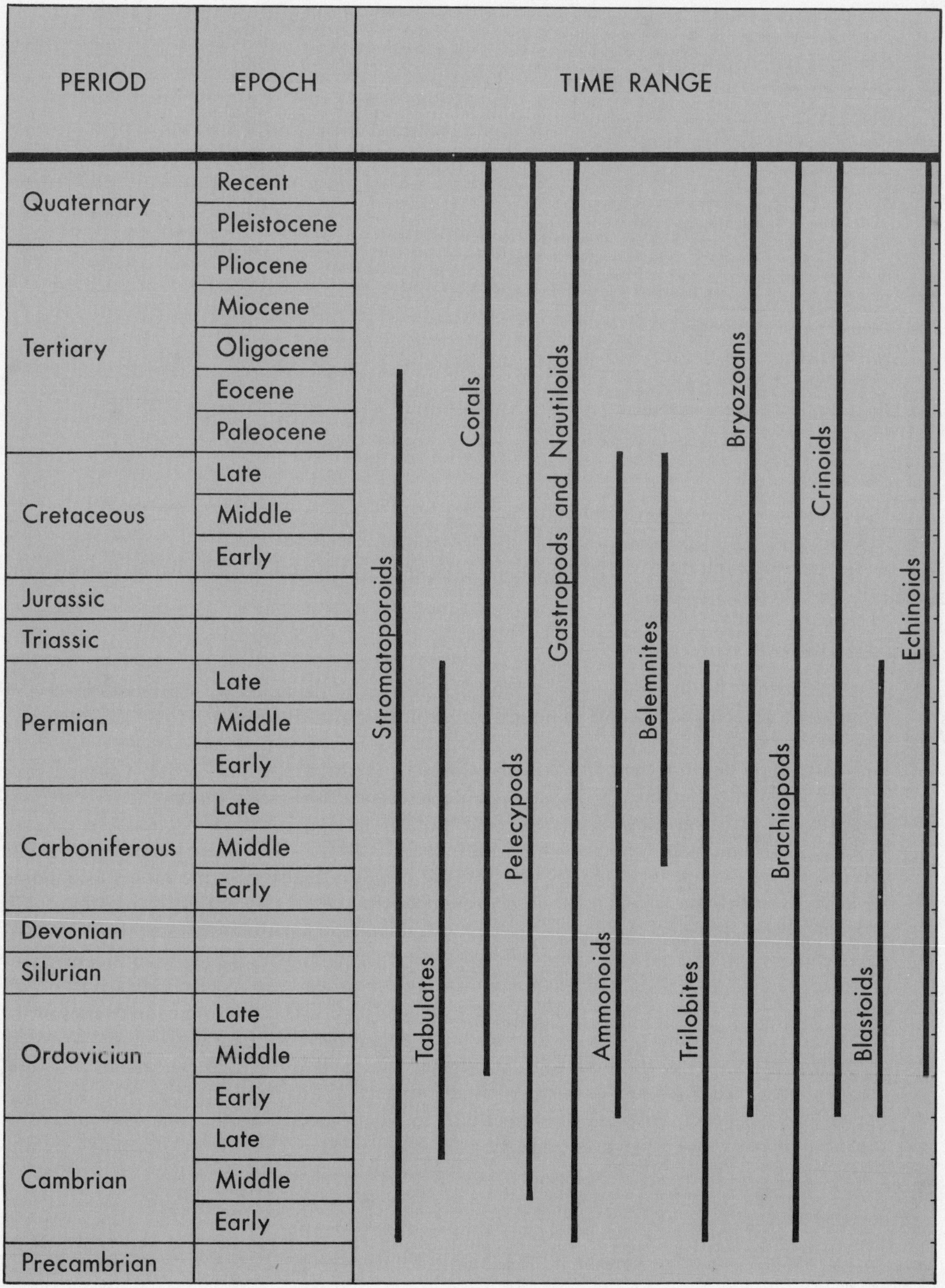
PERIOD
EPOCH
TIME RANGE
Quaternary
Recent
Pleistocene
Tertiary
Pliocene
Miocene
Oligocene
Eocene
Paleocene
Cretaceous
Late
Middle
Early
Jurassic
Triassic
Permian
Late
Middle
Early
Carboniferous
Late
Middle
Early
Devonian
Silurian
Ordovician
Late
Middle
Early
Cambrian
Late
Middle
Early
Precambrian
Stromatoporoids
Tabulates
Corals
Pelecypods
Gastropods and Nautiloids
Ammonoids
Belemnites
Trilobites
Bryozoans
Brachiopods
Crinoids
Blastoids
Echinoids

Figure 4.1 Time range of fossil groups studied in Chapter 3

Geologic Event	Age (before present) in years x 10^9
Origin of the earth	4.5
Widespread thermal event	3.6
Oldest organisms	3.2
Oldest photosynthetic organisms	2.0
End of banded iron deposition	1.8
Oldest eucaryotes	1.3
Oldest animals	0.65
Widespread animal life	0.60
Oldest vertebrates	0.47
Oldest land plants	0.40
Breakup of Pangea	0.22
Alpine (Laramide) orogeny	0.07
Onset of continental glaciation	0.002
Appearance of *Homo sapiens*	0.0005
Retreat of continental glaciation	0.000012
Beginning of written human history	0.000005

Part II

Using the overlapping range method and the time ranges given in Figure 4.1, identify as precisely as possible the age of each of the unknown slabs placed out in the laboratory. Each slab will contain several different fossil groups, and its age must be within the time interval in which the ranges of these groups overlap.

Test Questions

1. What procedures would be necessary to identify the ages of the unknown slabs more precisely? Why was it generally impossible to limit the time interval possible for each fossil assemblage to even a single geologic period?
2. Age correlation depends upon *a priori* knowledge of the time ranges of the fossil groups. How was this knowledge obtained, and what possible defects mar the use of this method?
3. Recently, several organisms that were presumed to have become extinct millions or even hundreds of millions of years ago have been found living in very deep water or otherwise inaccessible environments. Among these are the coelocanth fish and the monoplacophoran mollusks. The living sclerosponges closely resemble the presumably extinct stromatoporoids. What are the implications of these discoveries for geologic time correlation using fossils?
4. The genus *Hyolithellus*, a very common fossil in early Cambrian strata, is now considered to be a fossil representative of the Pogonophora, the bearded worms, close relatives of the vertebrates and a phylum once thought not to have a fossil record. Also the discoveries of the preservation of the soft tissues of animals lacking hard parts shows the existence of a fossil record for many groups that were long considered incapable of being preserved. In this light, are the presumed first appearances of fossil organisms any more reliable than their last appearances, or presumed times of extinction? To what extent does the time range of a fossil organism depend upon its preservability and abundance in certain rock-forming environments?

Background Reading

Berry, W. B. N., 1968, Growth of a Prehistoric Time Scale Based on Organic Evolution, W. H. Freeman and Co., San Francisco.

Eicher, D. L., 1968, Geologic Time, Prentice-Hall, Englewood Cliffs, N.J.

Harland, W. B., et al. (eds.), 1967, The Fossil Record, Geological Society, London.

Chapter Five

Paleoecology
Environmental Meaning of Fossil Types

Objectives

1. To evaluate different types of fossils in interpreting the physical conditions of ancient depositional environments.
2. To determine the physiological tolerances and preferences of living organisms commonly preserved as fossils.
3. To study how an organism's skeleton can react to external conditions during growth, and to use such responses in paleoecology.

Important Terms

Paleoecology—The study of the factors that control the distribution and abundance of fossil organisms in bodies of sedimentary rock.

Physiology—The set of functions that an organism carries out in order to live and reproduce, including food gathering, locomotion, digestion, respiration, and excretion.

Tolerance—The range of conditions which an organism can withstand and still survive.

Preference—A narrow range of conditions in which an organism functions best, both with respect to physical conditions and competition with other organisms.

Norm of reaction—The total variety of forms that a single organism can have, given different external conditions during growth.

Analogies with Living Counterparts

The best information concerning the tolerances and preferences of fossils normally comes from living counterparts that have similar physiologic requirements. Sometimes physiologic inferences can be made from well-preserved fossil skeletons alone, using mechanical arguments. The discussion below centers on turbidity and salinity requirements.

The amount of sediment suspended in the water restricts the distribution of many organisms (Figure 5.1). Many aquatic animals filter water through their organs to obtain food and dissolved

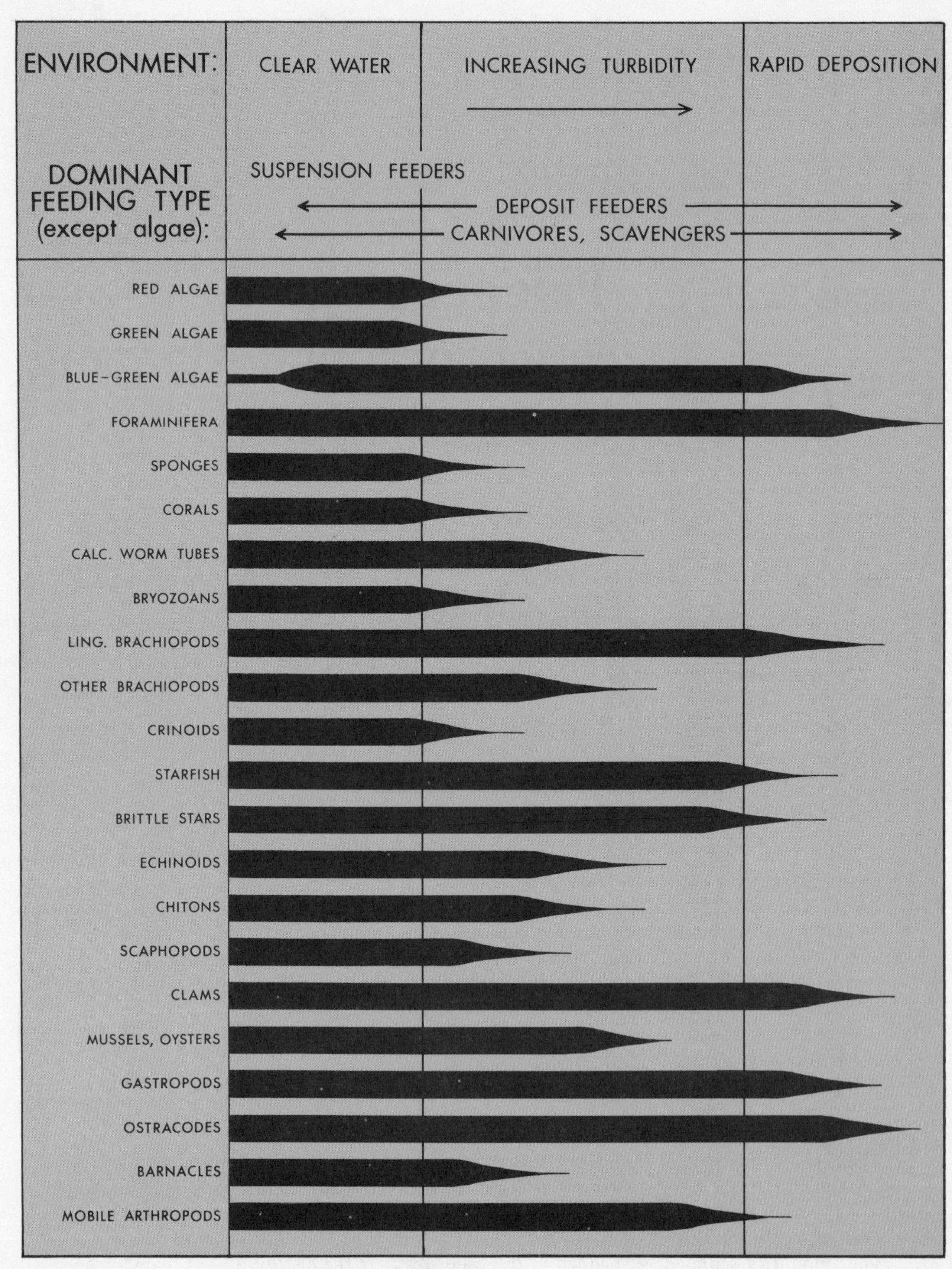

Figure 5.1 **Response of organic groups to turbidity and sedimentation**

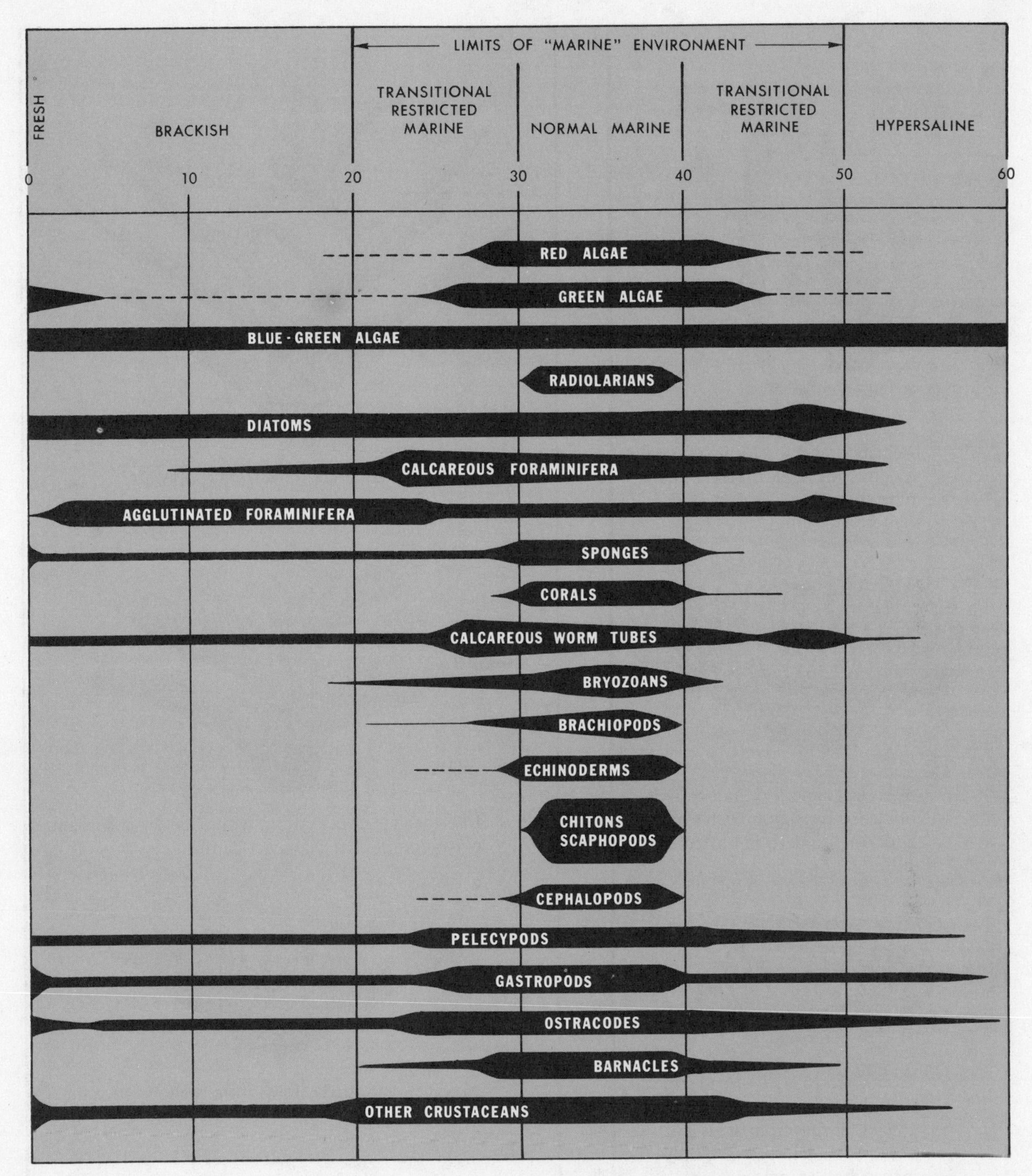

Figure 5.2 Response of organic groups to salinity; thickness of bar roughly reflects numbers of species. Numerical scale is measured in parts per thousand of dissolved salt.

oxygen and to dissipate waste materials. Many species have special adaptations, such as pore systems, sieve plates, meshes, and spines to keep out particles that are too large. Smaller particles, especially silt and clay, may clog the respiratory or feeding apparatus and cause death. If deposition is too rapid, certain organisms (especially attached forms) may not be able to grow or burrow up through the covering sediments quickly enough to prevent burial. Low, flat, thin shells or colonies are much more susceptible to burial than tall, erect, or branching forms. To summarize, response to

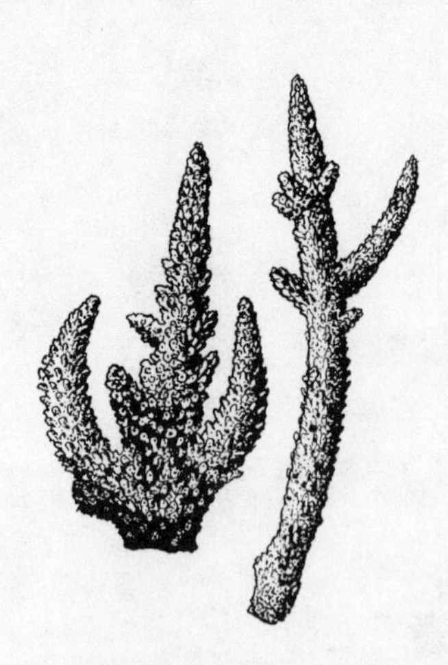

1. Delicate branching form from below low tide level, on coral knolls in protected lagoons and other sheltered localities.

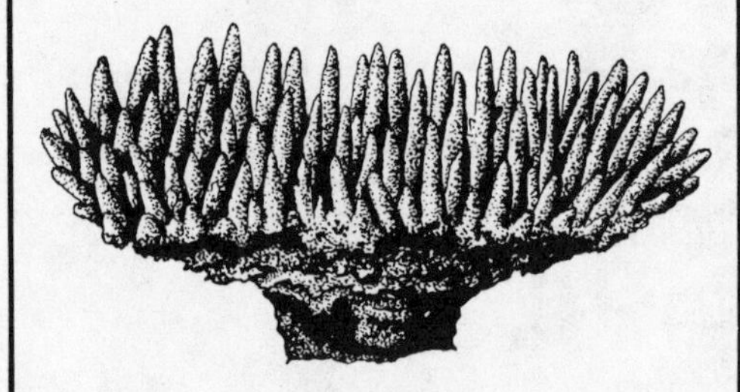

2. Shorter, stumpy, cylindrically branched form from water only a few meters deep in protected sites.

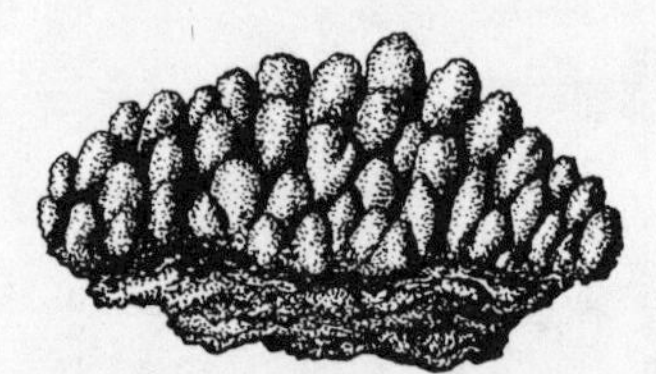

3. Broad, almost encrusting form with short, closely packed polygonal branches from areas exposed to rough water and only inches deep at low tide.

1. Tall, erect form from deep water.

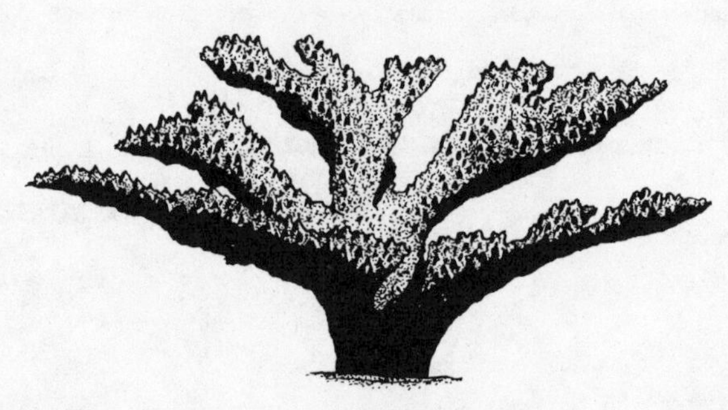

2. Intermediate, stocky form with conelike accordant crests common at shallower depths.

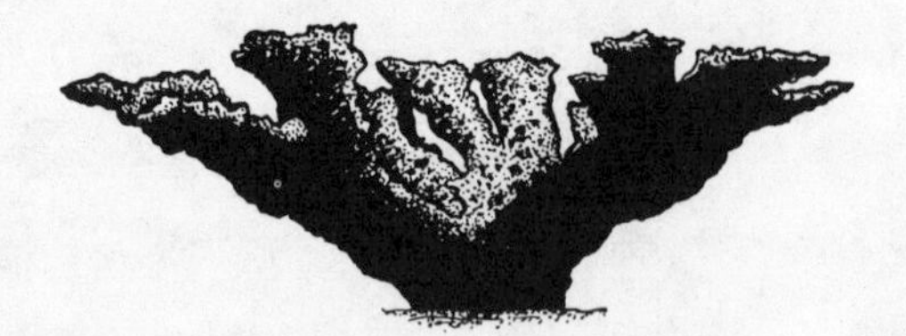

3. Massive, very stocky form from shallows near the reef crest.

A. Three growth forms of *Acropora humillis* from the Marshall Islands modified by differences in water turbulence and depth.

B. Three growth forms of *Acropora palmata* from Andros Island in the Bahamas modified by differences in water turbulence.

Figure 5.3 Environmental influence on the coral genus *Acropora*

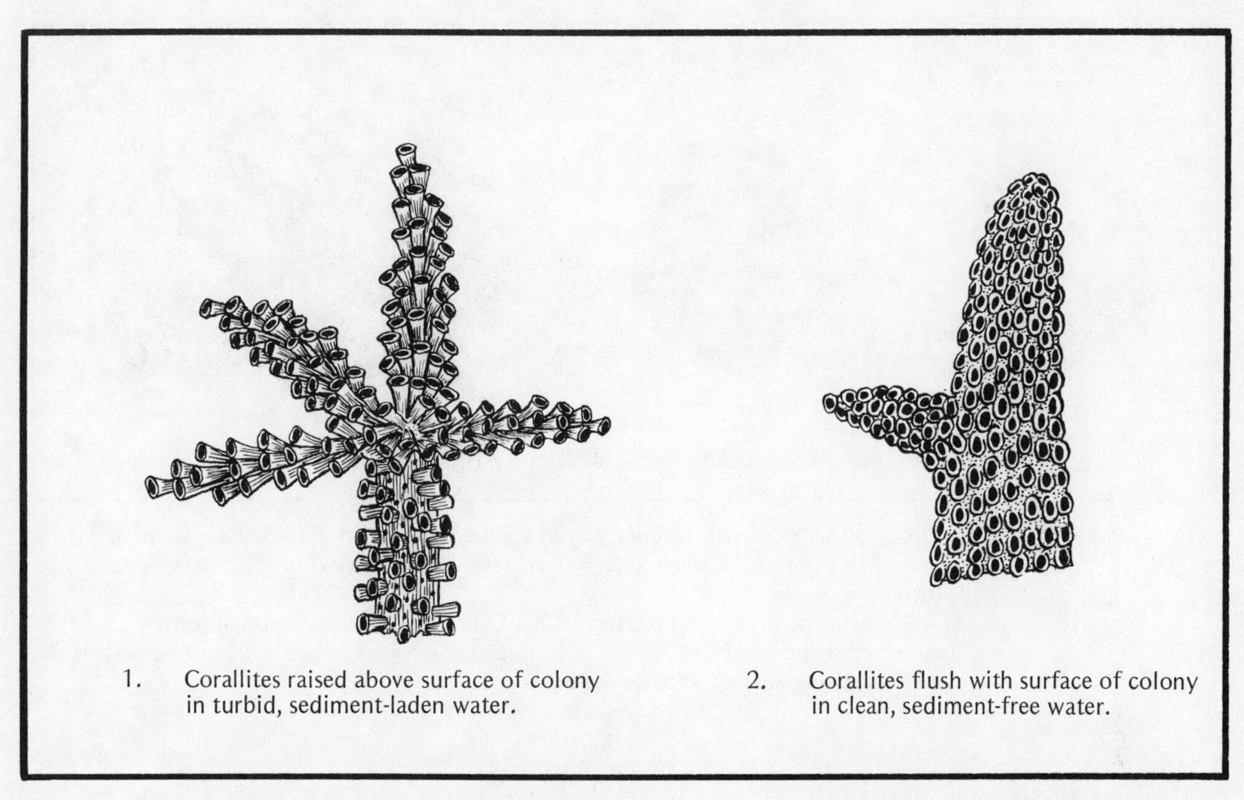

1. Corallites raised above surface of colony in turbid, sediment-laden water.

2. Corallites flush with surface of colony in clean, sediment-free water.

C. Differences in corallites of *Acropora pulchra* from the Pacific, demonstrating a response to sedimentation.

Figure 5.3 continued

turbidity and sedimentation can be determined from (a) living counterparts and (b) the design of the shell or skeleton.

The amount of dissolved salts in the water likewise restricts the distribution of many groups (Figure 5.2). Salinity is a useful guide to several major environments. *Fresh* water is found only on land masses. *Brackish* water is found in bays, lagoons, estuaries, deltas, and river mouths, all of which are shoreline environments. *Hypersaline* conditions characterize coastal lagoons and lakes in arid regions that have restricted circulation or no external drainage.

Calcareous-shelled brachiopods (articulates), stony bryozoans, corals, echinoderms, and cephalopods are good indicators of *marine* salinities (30.0 to 40.0 ‰ dissolved salt). In the echinoderms and corals, seawater serves as the major body fluid; consequently they cannot tolerate even slight salinity fluctuations. Abundant oysters or phosphatic brachiopods (inarticulates) are good indicators of brackish water (0.5 to 30.0 ‰), particularly if normal marine types are absent. Negative evidence is frequently important in establishing non-normal marine conditions. Some microorganisms may develop aberrant shell forms when grown in salinities that are abnormally high or low.

Norms of Reaction

Organic growth often follows different pathways when required to take place under different external conditions, particularly variations in water turbulence (agitation). Geneticists refer to such changes as the norm of reaction, or the amount of structural leeway that the genetic makeup of the organism will permit. Such variation might confuse the taxonomist, but it proves very useful to those who are looking for gauges on the ancient environment. Just as sediment grains respond to different degrees of water agitation, so do the shapes of many of the organisms living in the water.

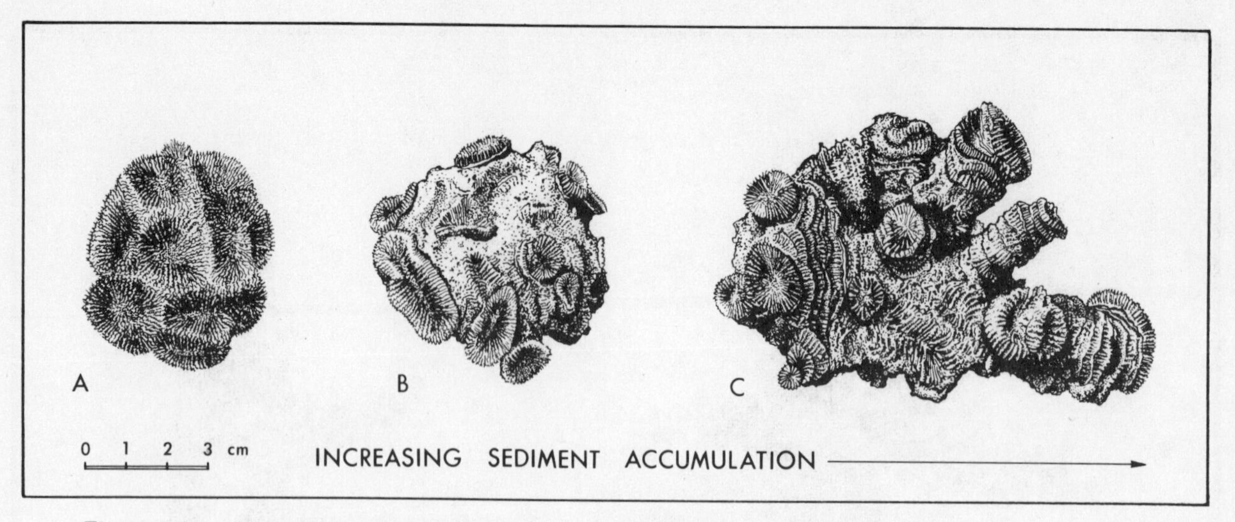

Figure 5.4 Variations in growth form of the coral ***Isophyllia multiflora*** **from Florida Bay, Florida.**
A. Specimen from a mud-free rocky platform showing "normal" development of colony form.
B. and C. Specimens from a carbonate mud bank showing effects on growth form with increased sediment accumulation. Columnar projections allow the coral to elevate its living tissue above encroaching mud.

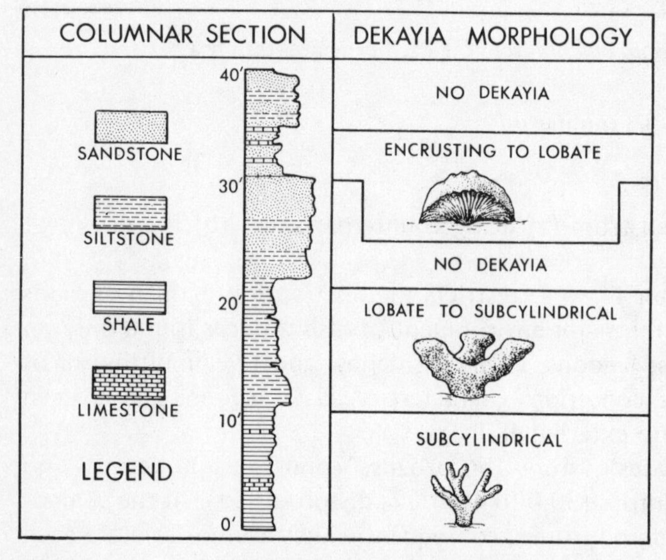

Figure 5.5 Changes in colony form in the bryozoan *Dekayia* associated with beds of differing lithology.

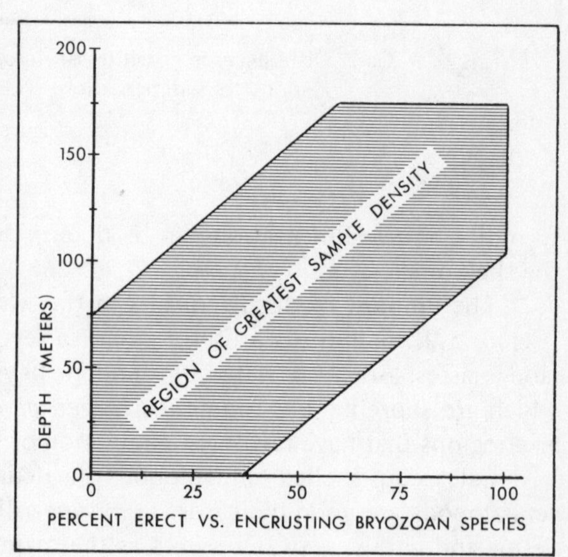

Figure 5.6 Increase of erect colony forms of bryozoan species with increasing water depth on the continental shelf off New England.

Reasons for such shape variation possibly include stresses on the growing animal, irritation of the shell-forming tissues, turbidity and sedimentation around the animal, and differences in the amount of food supplied and the flushing away of waste products.

Comparatively little is known about the norms of reaction of solitary organisms. Living brachiopods become wider and more ornamented in deep, quiet water, but fatter and smoother shelled in shallow, agitated water. Living mussels (clams) have shorter and rounder shells in rough water, and longer, more angular shells in quiet water. Both groups tend toward more spherical,

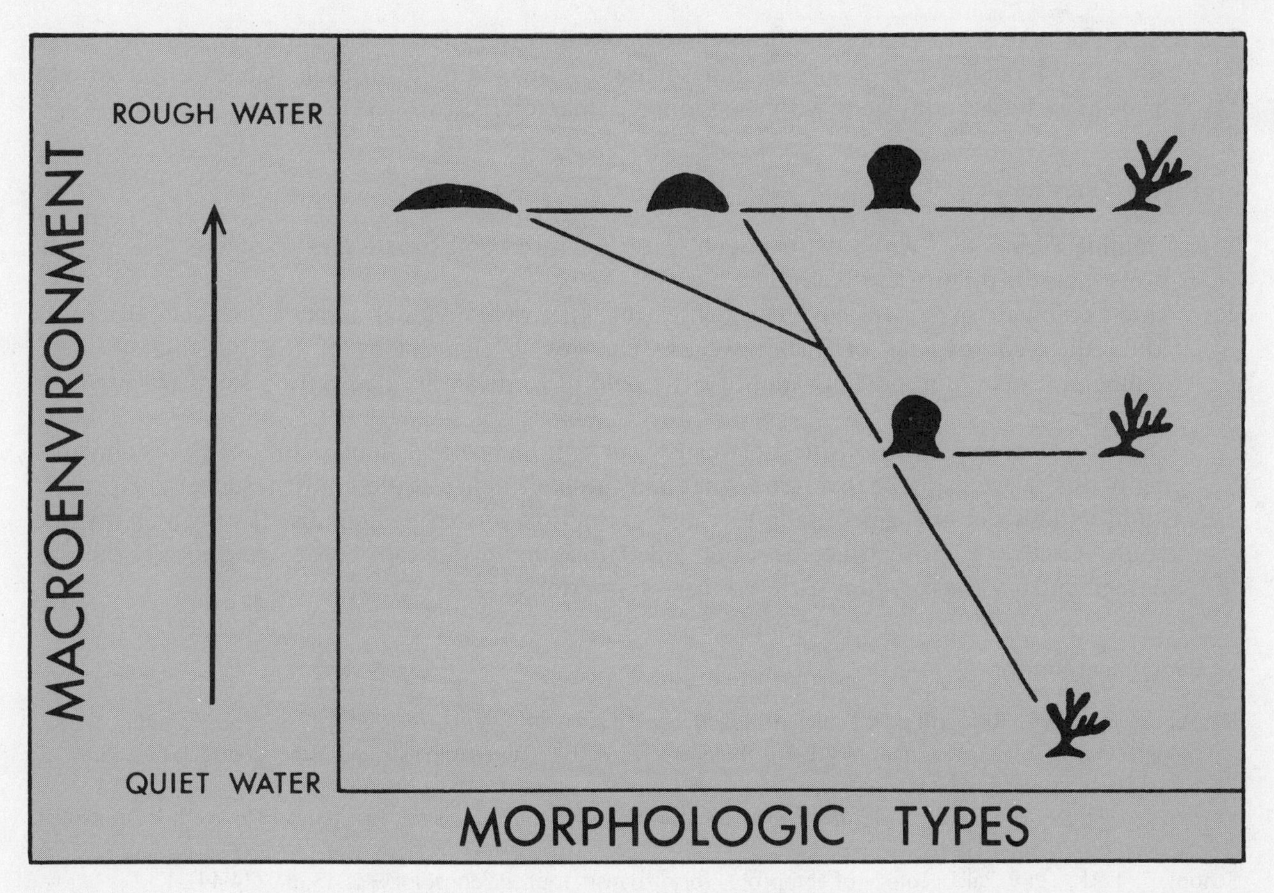

Figure 5.7 Growth forms of the encrusting protozoan *Homotrema* resulting from different degrees of water turbulence.

deep-bodied forms with thicker shells in more agitated conditions, and thinner, shallow-bodied forms with thinner shells in quiet water.

Conversely, much data has been published concerning the probable norms of reaction in colonial organisms (corals, tabulates, bryozoans, sponges, and stromatoporoids). Environmental effects on colony form as well as on individual members of the colony are shown for two modern corals (Figures 5.3, 5.4). In general, shorter, stockier forms inhabit shallower, more agitated water, and taller, branching forms live in deeper, quieter water. Lacy or fenestrate forms (sea fans and fenestrate bryozoans) live in a current; the openings allow water to move through the skeleton. The Ordovician bryozoan *Dekayia* (Figure 5.5) shows colonial forms similar to those of the living corals in beds of different lithology. The percentage of erect to encrusting forms generally increases with water depth in living species of bryozoans (Figure 5.6). Similar growth form changes are shown by the encrusting protozoan (foraminifer) *Homotrema rubrum* (Figure 5.7). Note that all growth forms are found in turbulent macrohabitats, but the rough water forms drop out as the water becomes quieter.

Exercise

Several large slabs bearing assemblages of fossils will be set out in the laboratory.

1. Using analogies with living counterparts, determine the turbidity and salinity of the macrohabitat that each slab represents. You may have to use the key from Chapter 3 to identify some of the fossils. Do any of the fossils have any special adaptations that might further specify the turbidity or deposition rate?

2. Using the norms of reaction of the shell and colony forms, determine the water turbulence for each slab. Examine the grain size and sorting of the sediment in each slab. Are the growth forms in each case consistent with the sediment characteristics?

Test Questions

1. Examine Figure 4.7. Why do the rough water environments have a greater diversity of growth forms than the quiet water habitats?
2. Which colonial forms will have the tightest packing of individual animals into the structure? Could different degrees of packing cause changes in the shapes of the individual colony members? Could you possibly identify the colony form if you had only a small fragment of the colony?
3. What is the general relationship between water turbulence and depth? In what environments might this relationship be inverted? What additional evidence could confirm such an inversion?
4. Could widespread changes in salinity cause extinction of marine animals? If so, which organic groups would be most affected? Find out, from any historical geology textbook, when in geologic history these groups suffered mass extinctions.

Background Reading

Heckel, P. H., 1972, Recognition of ancient shallow marine environments, *in* Rigby, J. K., and Hamblin, W. K., Recognition of Ancient Sedimentary Environments, Soc. Econ. Paleontologists and Mineralogists, Spec. Pub. 16, p. 226-86.

Lowenstam, H. A., 1967, Adaptive traits in skeletal morphology: American Geol. Inst., Short Course Lecture Notes, Paleoecology, p. HL1-HL13.

Schopf, T. J. M., 1969, Paleoecology of ectoprocts (bryozoans): Jour. Paleontology, v. 43, p. 234-44.

Chapter Six

Paleoecology
Fossil Communities

Objectives

1. To learn the characteristics of the major types of stable benthonic communities that have persisted through much of the Phanerozoic.
2. To consider the causes for the consistent association of particular plants and animals in an organic community.
3. To use these faunal associations to derive additional environmental information from ancient rock suites.

Important Terms

Community—A group of fossils that have a high degree of affinity and a consistent tendency to occur together.

Ecologic succession—A continuous sequence of communities that inhabit a particular area.

Pioneer—The initial community inhabiting a newly formed substrate.

Climax—The final or equilibrium community to inhabit a given area, after which no successional changes are observable unless external conditions are drastically changed.

Stable benthonic community—A recurrent association of functional or adaptive types of organisms which is independent of the presence of particular genera or species.

Organic Reef Communities

Organic reefs are major features of the marine environment that have existed throughout the Phanerozoic Eon. In the different habitats associated with the reef complex, different associations of organisms are found. Even though the particular organisms in each habitat have changed throughout time, their functional role in the habitat has not especially changed, and the same functional types appear again and again in these reef communities. The niches occupied by tabulate corals and stromatoporoids during the Paleozoic are now filled by scleractinean corals and calcareous algae. In this regard, the fossil record is marked with many examples of convergent evolution, in which organisms with different ancestries have assumed the same functional or adaptive role.

The major feature of a reef complex is the presence of a rigid organic framework that stands up above the seafloor, usually a linear feature fringing the margins of a landmass or island. Most reefs grow upward nearly to sea level, and create, through organic growth alone, a whole complex of

A. Pioneer community

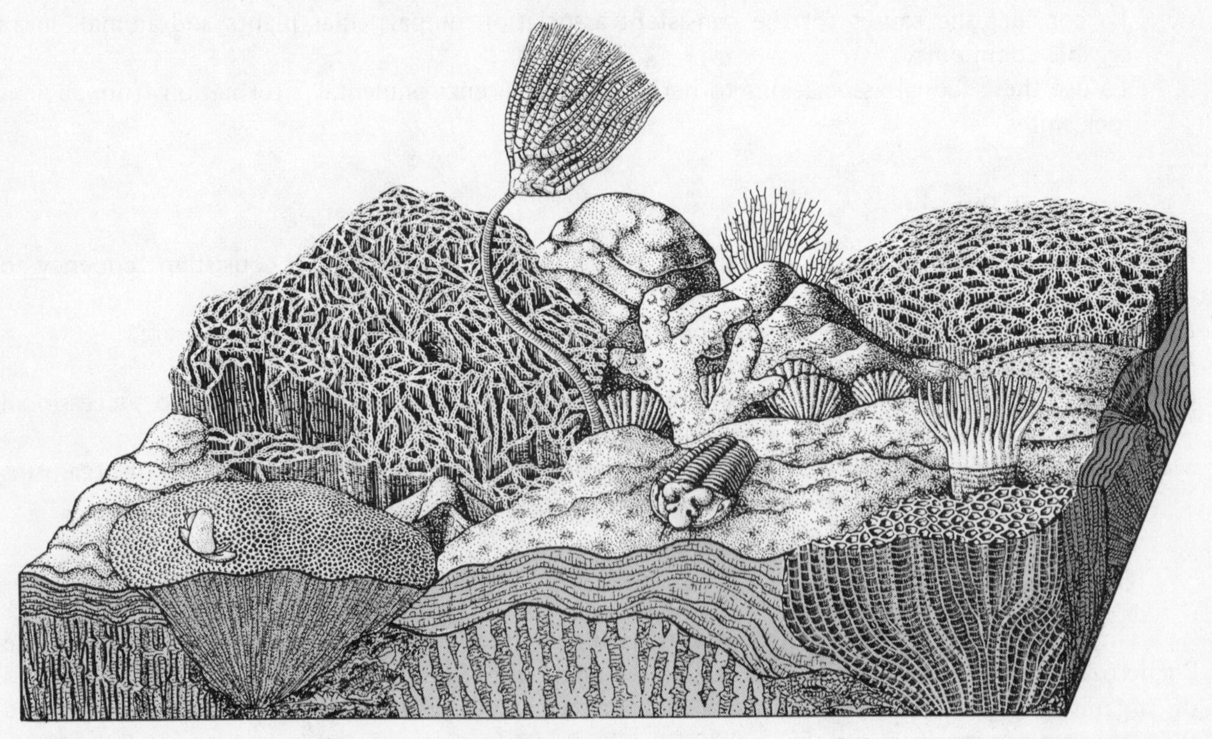

B. Intermediate community

Figure 6.1 **Succession of organic reef communities exemplified by the Middle Silurian in the Great Lakes region.**

C. Climax community (semiprotected niche)

Figure 6.1 continued

environments around them. The backreef lagoonal areas are characterized by quiet, shallow, clean, well-lighted waters, and may be bordered by extensive tidal flats landward. The reef core consists of the sediment-trapping, frame-building organisms, as well as the other organisms that inhabit the multitude of microhabitats within the core. The seaward flank of the reef is exposed to intense wave and current action. Three stages have been recognized in the ecologic succession of reef communities: a pioneer, quiet water stage; an intermediate, semirough water stage; and a climax, wave-resistant stage. The organisms that characterize each stage and their growth forms are shown in Figure 6.1. All three stages of succession exist contemporaneously in adjacent habitats; the pioneer community eventually inhabits the backreef lagoon. A series of Devonian communities and their associated sediment types are illustrated in Figure 6.2.

Detrital Substrate Communities

Ecologic succession has never been observed in communities inhabiting detrital sediment substrates. Such communities might better fit the definition above of stable benthonic communities (or recurrent communities). Stable communities are characteristically dominated, in terms of biomass, by only one of the feeding types described in Chapter 3. The Paleozoic Era seems to be

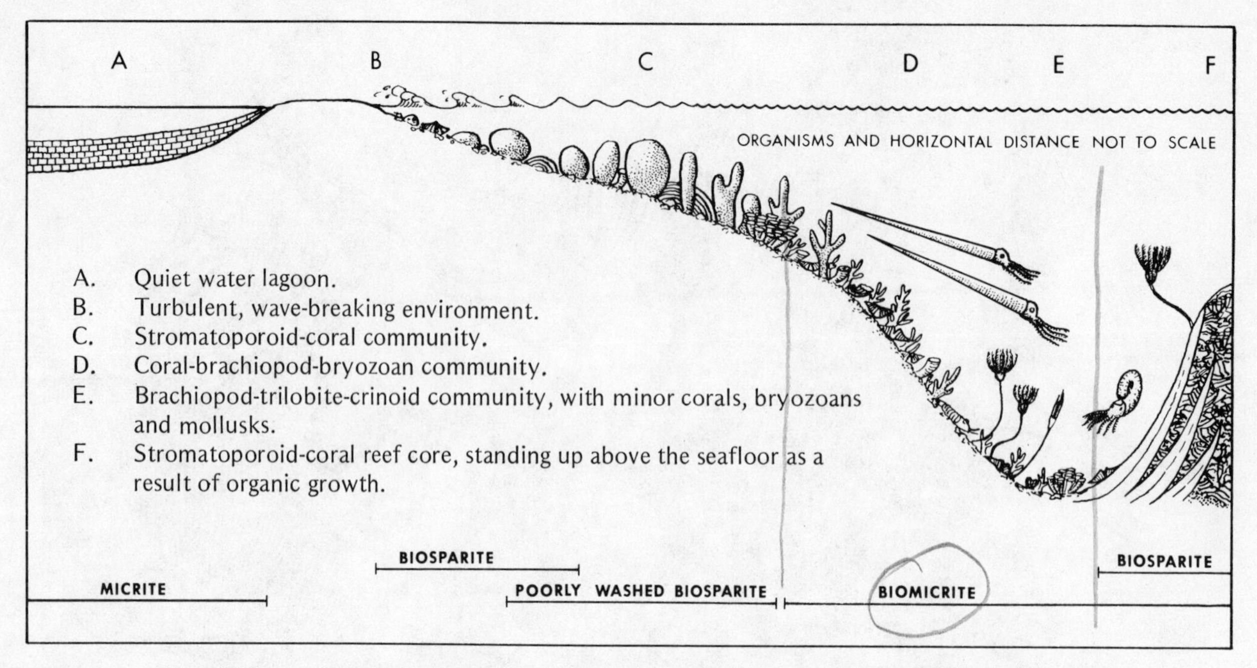

Figure 6.2 **Contemporaneous organic reef communities from the Devonian of Michigan**

characterized by three major detrital substrate communities: (1) an onshore assemblage of phosphatic brachiopods, gastropods, and pelecypods; (2) an intermediate shallow shelf assemblage of thin, smooth-shelled or spherical, spinose brachiopods and trilobites; and (3) an offshore (outer shelf) assemblage of spherical, thick-shelled brachiopods, thinly branching bryozoans, and crinoids (Figure 6.3). The onshore community seems to reflect brackish, coastal waters; the intermediate community, shallow, subtidal nearshore conditions; the offshore community, deeper, quieter water on the outer shelf, probably below wave base.

Exercise

A large collection of slabs representing diverse fossil assemblages will be placed out in the laboratory.

1. What is the dominant feeding type, in terms of skeletal volume, present in each slab?
2. To which group of communities does each slab belong, and which depositional environment does it represent?
3. Make a chart listing all of the types of fossils present in all of the slabs. List each fossil type both along the top and along the left margin of the chart, as shown:

	TYPE 1	TYPE 2	TYPE 3	. . . . .
TYPE 2				
TYPE 3				
. . . . . .				

Score the number of times that different types of fossils occur together in the same slab. Do you find the same associational groups as recognized in the communities outlined above?

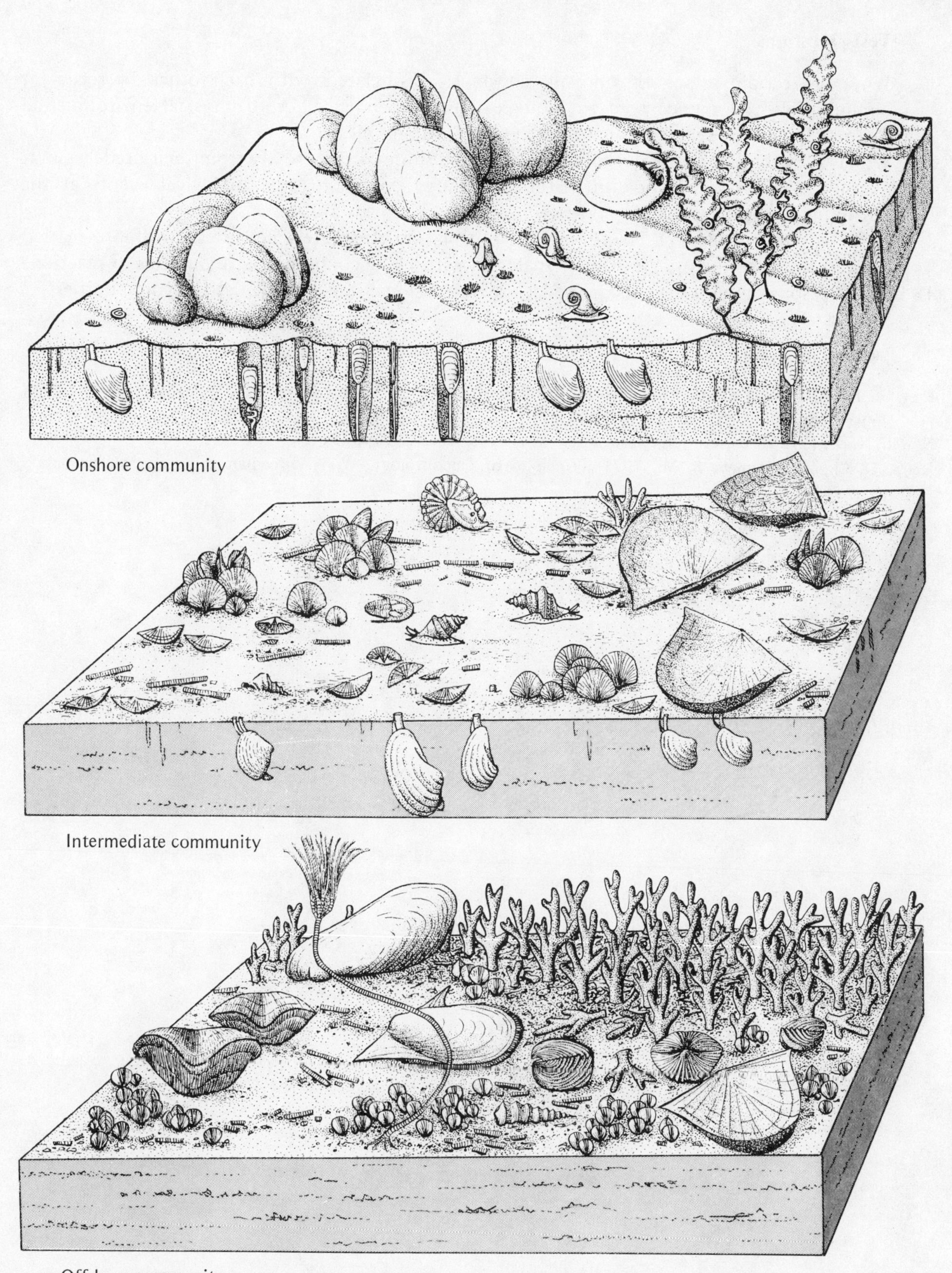

Figure 6.3 Major types of recurrent detrital substrate communities exemplified by the Appalachian Late Ordovician.

Test Questions

1. Why are certain types of animals constantly associated with each other in recurrent communities? Is it due more to biological interdependence, or to physical environmental factors? Would the same be true for successional communities?
2. Why is it likely that post-Paleozoic bottom communities differ greatly from their predecessors? Consult a general textbook in historical geology concerning the biological events at the beginning of the Mesozoic.
3. Is the community-derived environmental information completely consistent with information derived from sediments, life habits, analogies with living counterparts, and norms of reaction?
4. What reasons can you suggest for the lack of successional stages in the detrital communities?

Background Reading

Bretsky, P. W., 1969, Evolution of Paleozoic benthic marine invertebrate communities: Palaeogeography, Palaeoclimatology, Palaeoecology, v. 6, p. 45-59.

Newell, N. D., 1972, The evolution of reefs: Scientific American, v. 227, p. 54-65.

Raup, D. M., and Stanley, S. M., 1971, Principles of Paleontology, W. H. Freeman and Co., San Francisco, p. 193-254.

Chapter Seven

Sedimentary Petrology Grains and Matrix of Carbonate Rocks

Objectives

1. To understand the origin of the grains, matrix, and textures of carbonate rocks.
2. To relate the textural characteristics of the different rock types to environmental factors, such as distance from shore, level of water energy affecting the substrate, salinity, plant growth, and other biotic activity.

Important Terms

Calcite—Mineral form of $CaCO_3$ having a rhombohedral crystal lattice.

Aragonite—Mineral form of $CaCO_3$ having an orthorhombic crystal lattice.

Dolomite—Calcium magnesium carbonate, $CaMg(CO_3)_2$.

Limestone—Rock composed of more than 50 percent aragonite and/or calcite by weight.

Dolostone—Rock composed primarily of dolomite.

Allochems—The grains of a carbonate rock, analogous to grains of quartz or pebbles in a detrital rock; may be fossils, oolites, intraclasts, or pellets (see below).

Matrix—The silt and clay-sized fraction of a carbonate rock (*micrite*, or lithified lime mud), or clear calcite cement (*spar*) filling in void spaces between grains.

Oolites—Small spherical sand-sized (or slightly larger) grains; each has a nucleus surrounded by concentric layers or radially oriented needles of calcite or aragonite; formed by continuous rolling in highly agitated water supersaturated with calcium carbonate to coat the grains; assymmetrical oolites can form in quiet water conditions.

Intraclasts—Eroded lithified remnants of a preexisting rock or semiconsolidated sediment, usually a carbonate; examples are the flat pebbles in a *flat pebble conglomerate* and the reworked clusters of lithified oolites or pellets called "grapestone."

Pellets—Round sand-sized (or larger) grains having no internal structure, but composed of homogeneous lime mud; externally resembling oolites; may be fecal material produced by animals that ingest carbonate sediment to extract its organic content as food.

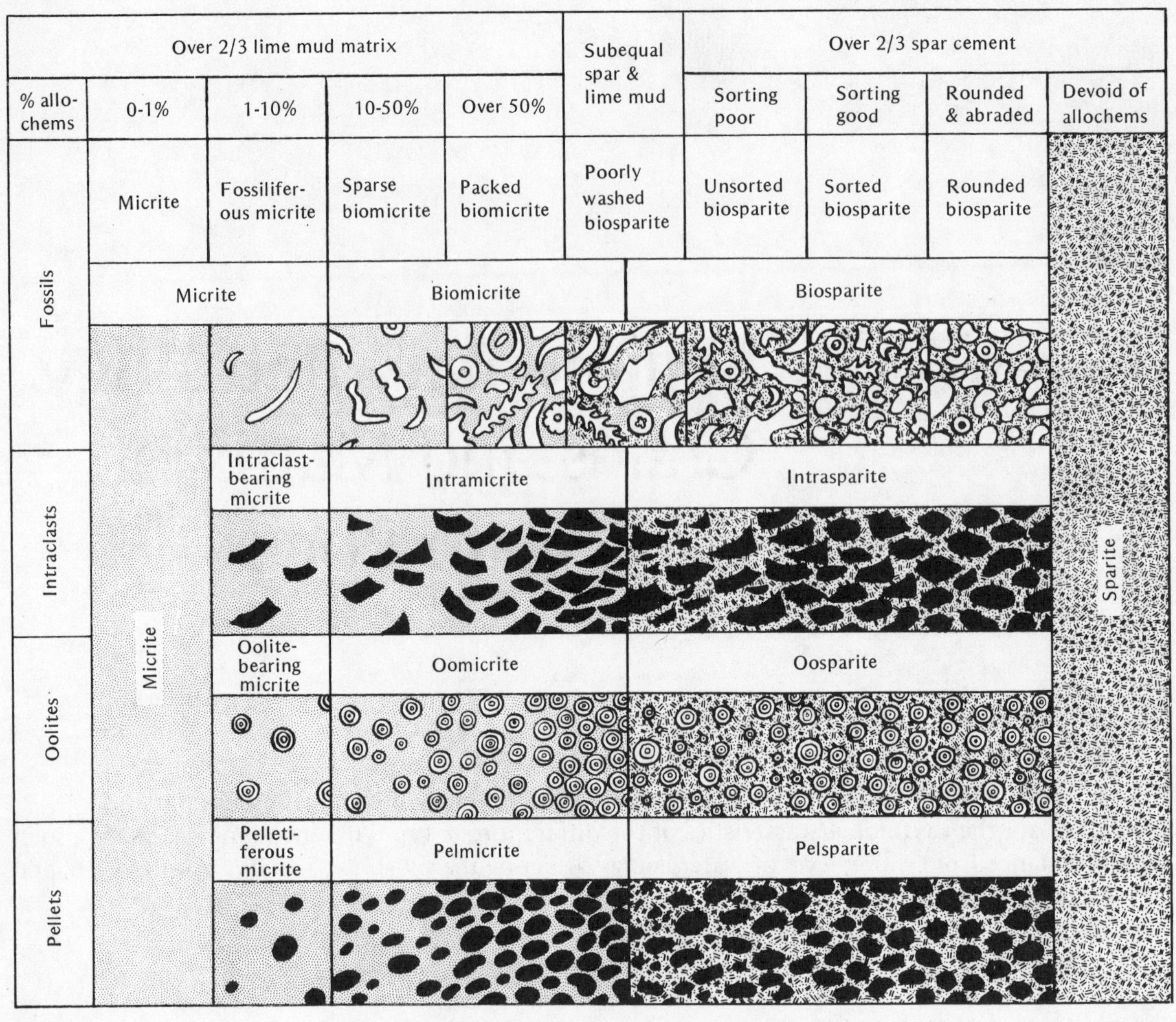

Figure 7.1 A classification of limestone rock types

EXAMPLES OF COMBINATIONS

Fossiliferous Intrasparite
Oolitic Pelmicrite
Pelletiferous Oosparite
Intraclastic Biomicrite

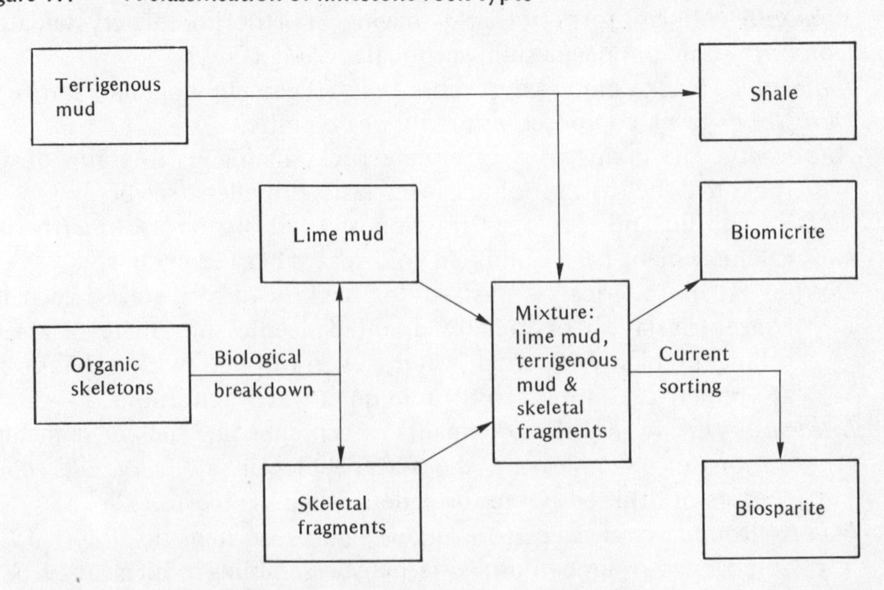

Figure 7.2 A simplified model for the origin of carbonate rock types involving input of terrigenous (land-derived) detritus.

Discussion

Carbonate rocks are formed from sediment that originates at or very near the site of deposition. Detrital rocks, in contrast, are composed of materials that originated at considerable distance and have been transported to the site of deposition. Thus detrital rocks provide information about the source area and transport processes as well as the environment of deposition; carbonates provide information primarily about the depositional environment.

Because much original carbonate is calcite, unaltered skeletal fragments suggest that the rock is primarily calcite in composition. Aragonite has a different crystal lattice, and is easily converted to calcite or dolomite; it is rare in most ancient rocks, especially those older than the Cenozoic. Dolomite forms in shallow intertidal or supratidal environments, mostly from preexisting aragonite skeletal debris in contact with seawater. Dolostones are often "sugary" or granular in texture, very porous, and contain recrystallized or altered fossil fragments.

Carbonate sediments are produced in almost all aquatic environments, especially those where plant and animal life is prolific. Wherever streams or currents supply detritus, any carbonate present is heavily diluted and then occurs as random grains in a detrital rock. Pure or undiluted carbonates are most common along coastlines or shelf areas where little, if any, detritus is available. The only places in North America where undiluted carbonates are forming today are the Florida Keys and the Bahamas. Both areas are adequately removed from any source of detrital sediment.

The grains of carbonate rocks are called allochems, and include fossil fragments, oolites, intraclasts, and pellets. The matrix is either lime mud, which lithifies to micrite, or clear, sparry calcite cement (spar). Spar has distinct crystal boundaries, and often results from the recrystallization of micrite. Micrite resembles detrital mud or silt in appearance, whereas spar often shows the cleavage faces of small calcite crystals. Be careful not to confuse the calcite cleavages in small fossil fragments, such as crinoid stems, with sparry cement. Thus the two types of matrix and the four types of allochems produce eight types of limestone (Figure 7.1). The processes that generate the principal limestone types on a mixed carbonate-detrital shelf are shown in Figure 7.2.

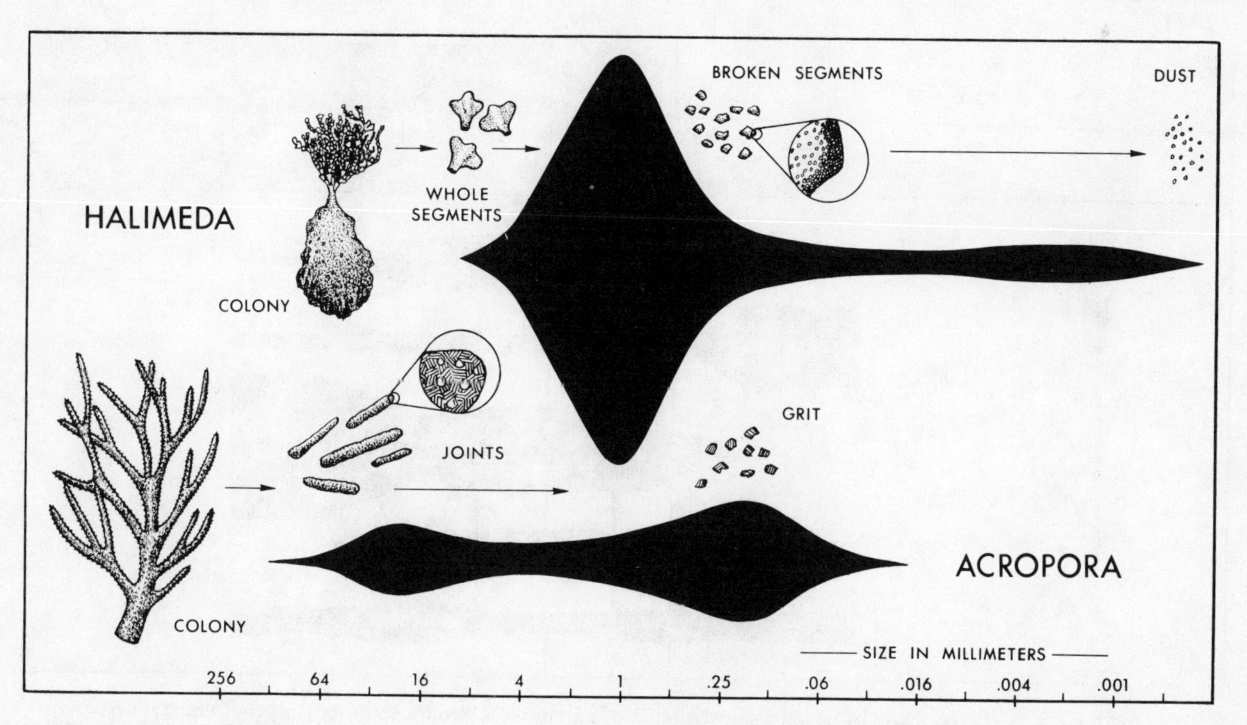

Figure 7.3 Breakdown of a coral *Acropora* and a green alga *Halimeda* into bimodal carbonate sediments

Organic skeletons most commonly disintegrate into a sediment that has a bimodal size distribution (Figure 7.3). The larger skeletal fragments (to the left) form the grains in a carbonate rock, whereas the grit or dust resulting from the breakdown of the skeletal microstructure forms the

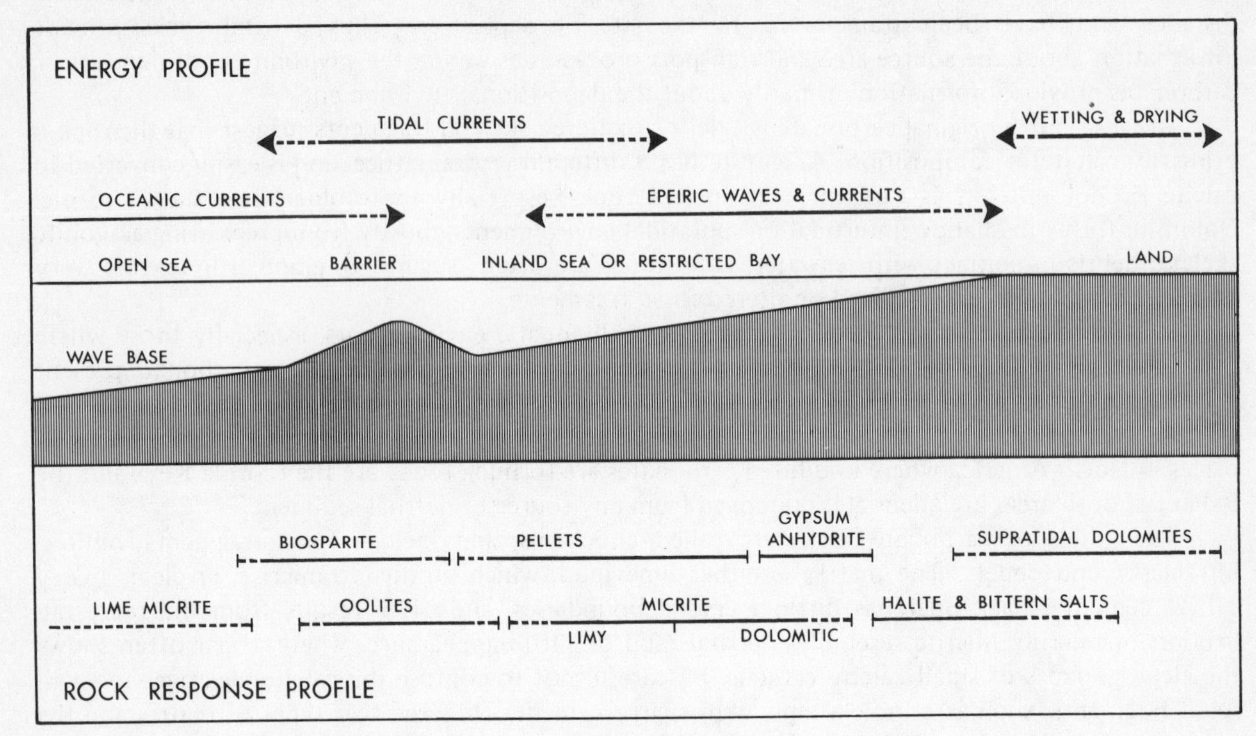

Figure 7.4 Simplified energy and rock response profile across a stable shelf having no input of land-derived detritus.

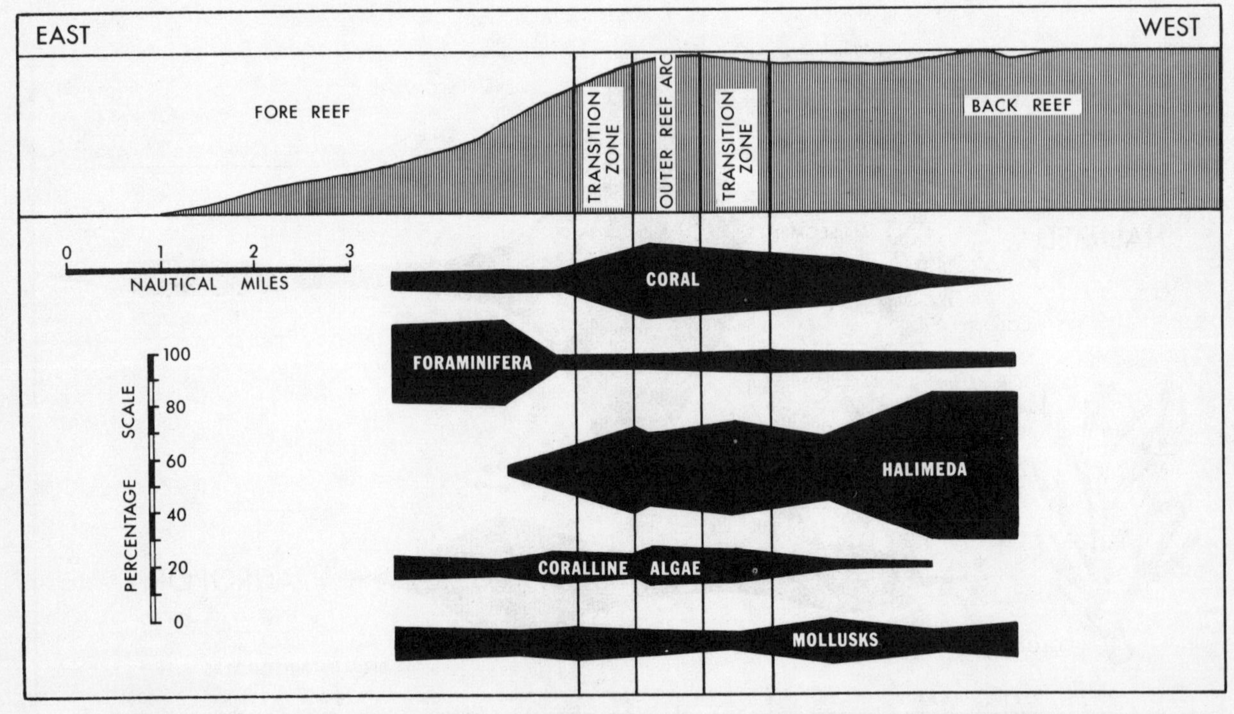

Figure 7.5 Percentages of organic constituents of calcium carbonate sediments in the reef tract of southern Florida.

lime mud matrix. Green algae (Figures 7.5, 7.6) are the principal source of lime mud in many modern carbonate environments. Algae and other plants tend to stabilize the fine-grained lime mud even in environments having high water agitation (Figure 7.6), thus contradicting the simple energy profile (Figure 7.4) as a guide to rock type.

Effects of Water Agitation

Spar represents the filling of void space in a coarsely grained carbonate sediment by secondary calcite cement, which suggests that the original lime mud, the product of skeletal disintegration, has been removed by water agitation. Thus rocks containing less lime mud and more spar might be considered to represent more agitated conditions in which the lime mud has been winnowed away into either quiet water lagoons or deeper parts of the shelf not affected by wave action (Figures 7.1, 7.2, 7.3, 7.4). The Great Bahama Bank (Figure 7.7) illustrates this process: an elongate oolite bar lines the western margin of the bank, where water agitation is highest; fragmented skeletal debris and intraclasts, free of lime mud, line the northern and southern margins of the bank. The central lagoon, having the quietest water, has the highest accumulation of lime mud. Organic growth is also highest in the lagoon, and fine-grained sediment is stabilized by plant growth, mainly turtle grass.

Biotic Effects

Conclusions drawn from energy processes must always be modified because of the effects of organic life on carbonate sediments. Although the rounding and sorting of shell fragments appear superficially analogous to those of detrital sediments, it is misleading to interpret carbonate textures solely in terms of hydrodynamics. Disintegration of organic skeletons, however, can take place entirely in quiet water by predominantly biological processes. The disintegration of fragile and semiresistant skeletons, such as those of algae and bryozoans, has probably served as a major source of many ancient micrites. Boring by algae and larger organisms into calcareous skeletons not only can produce lime mud by itself, but also serves to weaken the skeleton and expedite further skeletal breakdown. Bacteria can cause decay of the organic matrix that holds many skeletons intact. Sediment-ingesting organisms, such as sea cucumbers, may cause the sorting out of particles too large to ingest, and may cause rounding of carbonate grains within their digestive tracts. Thus, sorting and rounding can be produced biotically, with no action by water being needed. Mangrove roots, turtle grass, green algae, and algal mats (Chapter 10) can entrap lime mud, even in highly agitated water. The meshlike fenestrate bryozoans provided baffles to entrap lime mud in the core sediments of many ancient reefs.

Exercise

Two demonstrations will be set up in the laboratory to illustrate map views and cross sections of (1) a coral reef tract (such as the Florida Keys), and (2) an oolite bar-carbonate island (such as the Great Bahama Bank). Topographic and bathymetric maps of such areas will also be available. Examples of all of the major types of carbonate rocks will be keyed into their places of origin in the demonstrations.

Each student will be given a set of unknown carbonate rocks. Using a binocular microscope or hand lens, the following should be determined for each unknown:

a. all grain types (allochems) present
b. relative proportions of spar and micrite
c. rock name
d. fossil types identifiable from skeletal fragments

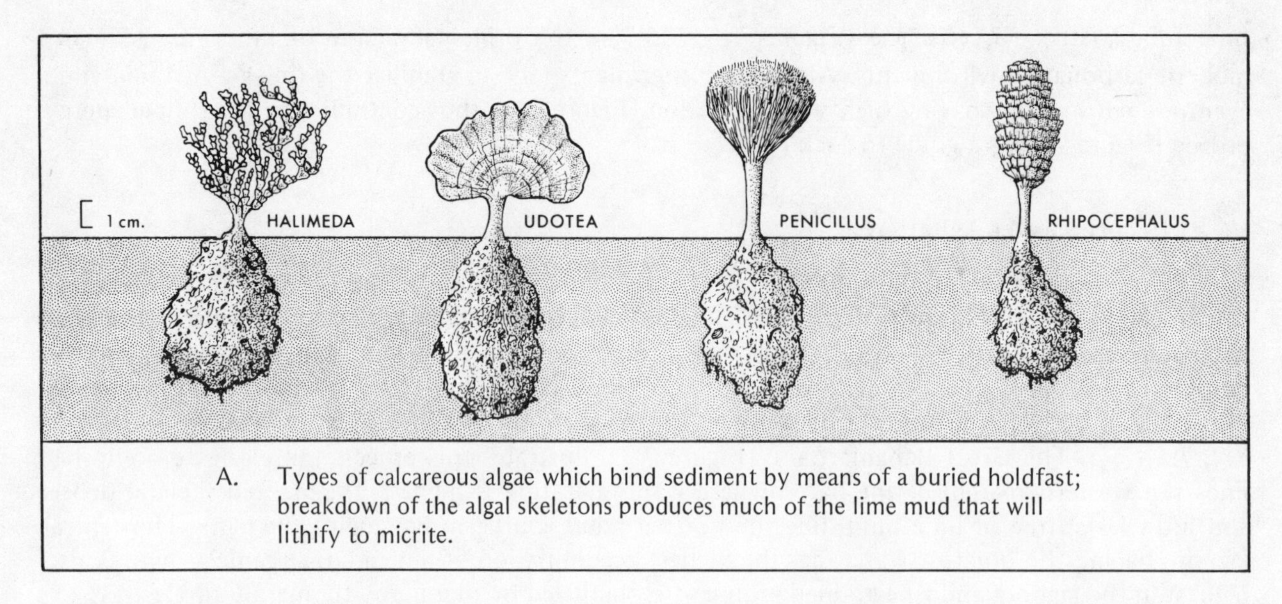

A. Types of calcareous algae which bind sediment by means of a buried holdfast; breakdown of the algal skeletons produces much of the lime mud that will lithify to micrite.

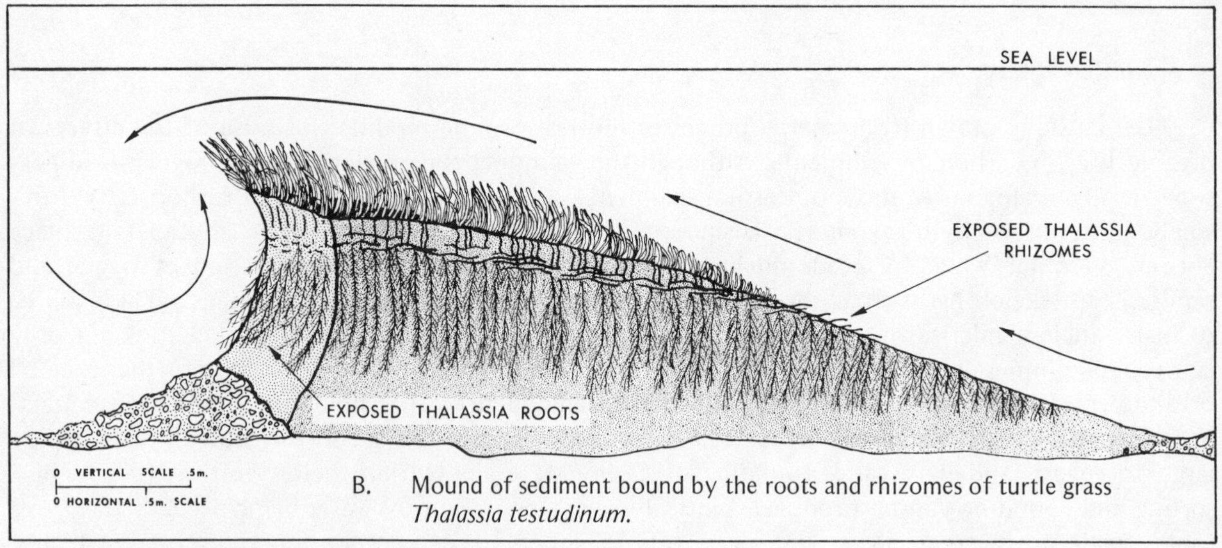

B. Mound of sediment bound by the roots and rhizomes of turtle grass *Thalassia testudinum.*

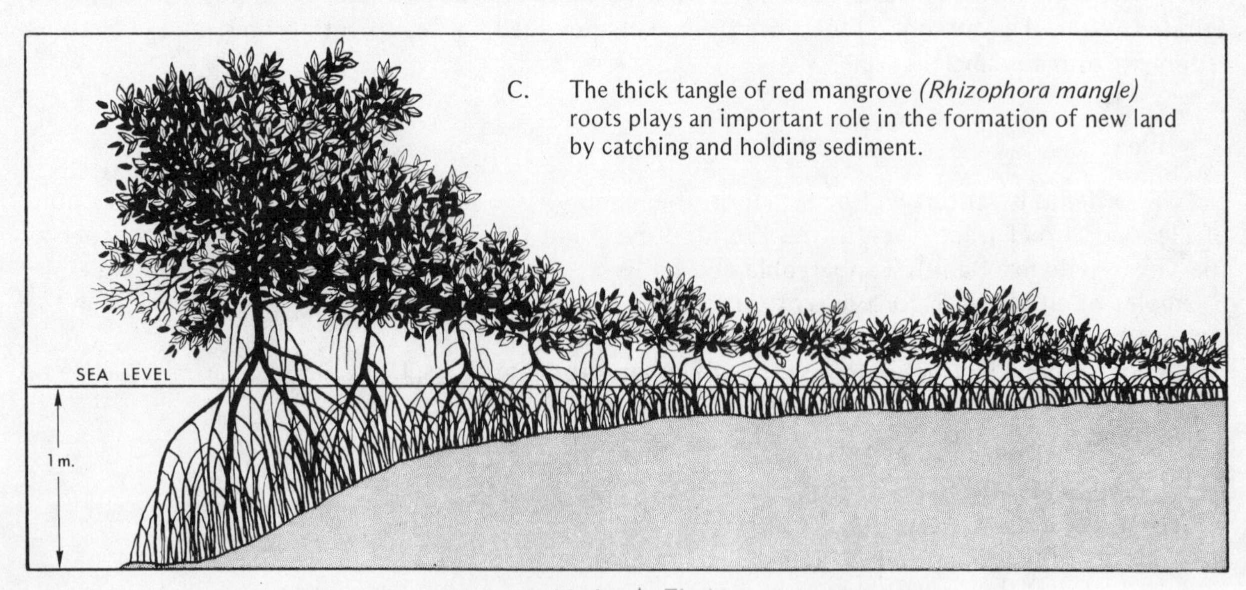

C. The thick tangle of red mangrove *(Rhizophora mangle)* roots plays an important role in the formation of new land by catching and holding sediment.

Figure 7.6 Sediment binders in Florida and the West Indies

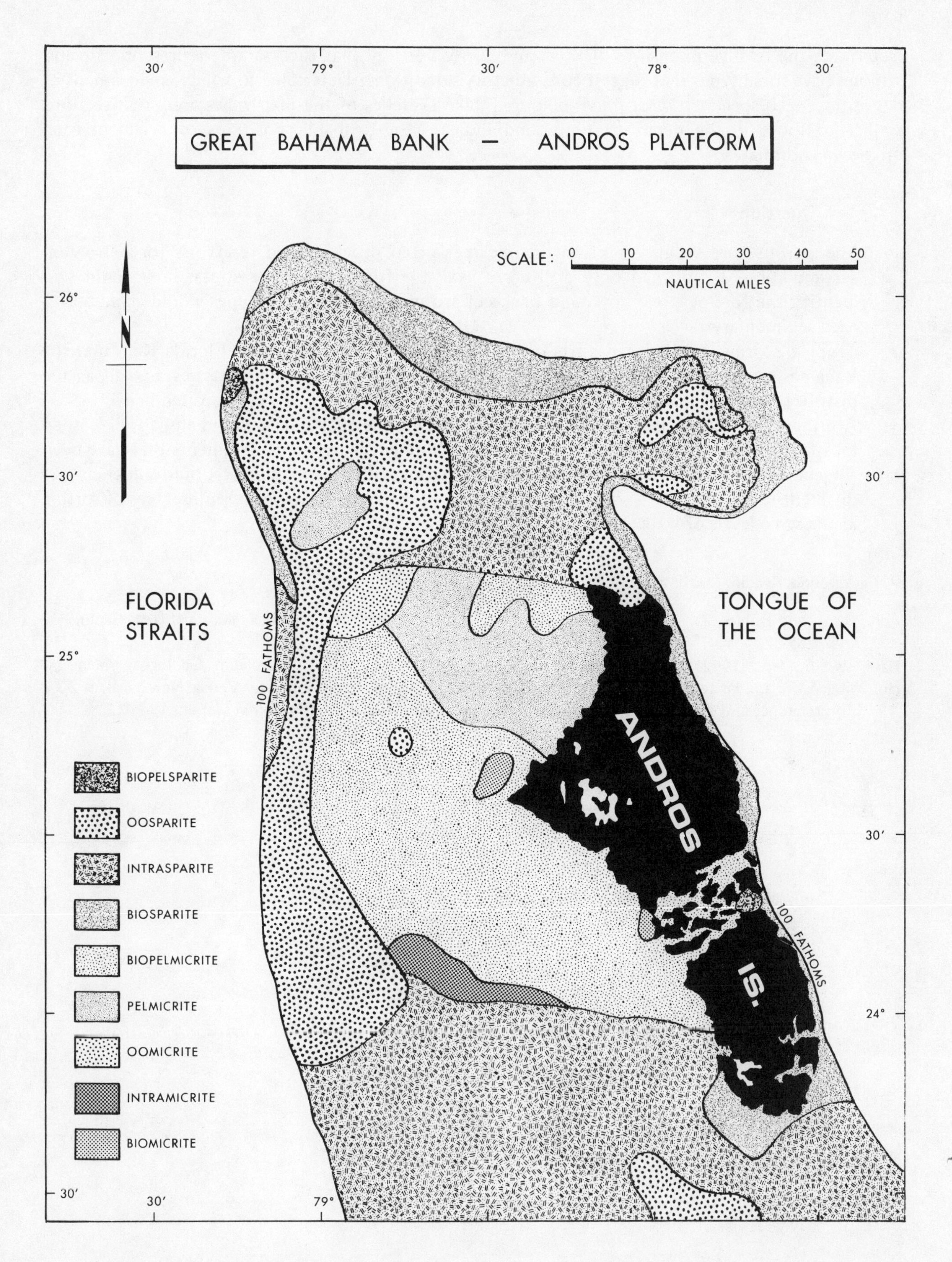

Figure 7.7 Map of contemporary carbonate sediments accumulating on the Great Bahama Bank

Determine the relative intensities of wave and current energy that each sample represents. Do any samples have fossil types that suggest contradictory conditions? Determine to which environment of deposition each specimen *most likely* belongs. Make sketches of the map views and cross sections similar to those in the demonstrations, and indicate the probable positions of origin of each unknown carbonate.

Test Questions

1. The porous core sediments of an organic reef often provide good reservoirs for petroleum accumulation. If you had only drill cores available from subsurface strata, how could you identify the forereef, reef core, and backreef sediments? What fossil types would characterize each sedimentary facies?
2. Tropical storms and hurricanes frequently disrupt sedimentation in the Florida Keys and the Bahamas every year. What effects will they have on carbonate rock types as well as the distribution of fossil shells? How could you recognize an ancient storm-deposited bed?
3. Are there necessarily any textural differences between a quartz beach sand and a well-washed biosparite? Between a detrital mudstone and a biomicrite? Will shell fragments that have been subject to high water energy have different shapes than sand grains? Could allochems of different shape affect the sorting or winnowing process, producing different degrees of sorting at the same levels of water energy?

Background Reading

Folk, R. L., and Robles, R., 1964, Carbonate sands of Isla Perez, Alacran reef complex, Yucatan: Jour. Geology, v. 72, p. 255-92.

Ham, W. E., ed., 1962, Classification of Carbonate Rocks, Am. Assoc. Petroleum Geologists, Mem. 1.

Horowitz, A. S., and Potter, P. E., 1971, Introductory Petrography of Fossils, Springer-Verlag, New York, p. 2-33. (This reference also contains 100 plates showing the details of fossil allochems in thin sections.)

Chapter Eight

Sedimentology
Inorganic Sedimentary Structures

Objectives

1. To recognize the geometry of homogeneous grain aggregates in sedimentary rocks that are largely the result of nonorganic processes.
2. To understand the physical processes that produced such structures.
3. To interpret the shape and distribution of inorganic sedimentary structures in terms of their environments of origin.

Important Terms

Bedding—Planar differentiation of sediment into homogeneous layered units, which range in thickness from a few millimeters to several meters. Thinner stratification is termed *lamination*. Cyclic repetitions of fine and coarse grained laminations are *varves*.

Cross bedding—Nonhorizontal bedding, caused by initial deposition of sediment on an inclined surface. Sediment deposited in a ripple ridge or sand dune forms cross beds or cross laminae. Usually the direction of inclination reflects the sense of movement of the transporting medium.

Ripple marks—Parallel ridges of sediment perpendicular to the direction of movement of the transporting medium. Changing current directions may superimpose a second set of ripples on top of a former set, producing rhombic interference ripple marks. The wave length and amplitude of ripple marks reflect the wave forms of the transporting medium.

Sole marks—Directional markings on the lower side of a bed parallel to the direction of water movement.

Desiccation cracks—V-shaped cracks that subdivide a sediment layer into polygonal plates. The hardening, erosion, and redeposition of these plates produce a *flat pebble conglomerate.* These plates may also be shingled up in an imbricate structure, producing an "edgewise" conglomerate.

Raindrop prints—Along with mudcracks, raindrop and hailstone imprints reflect exposure of soft, unconsolidated sediment to the air.

Discussion

Sedimentary structures primarily result from relatively straightforward physical processes. Horizontal bedding implies deposition on a horizontal surface, and a low angle of repose of the sediment particles. Angular grains tend to interlock; round grains do not interlock, resulting in beds of lower inclination. The depositing medium plays an important role: the same sediment will have more steeply inclined bedding in air than in water. What effect does water have on the steepness of bedding? Cyclic bedding implies periodic pulses of sediment associated with tides, spring floods, yearly turnover in lakes, and other cyclic phenomena that control sedimentation. Inclined bedding reflects the directional movement of wind or water. If cross beds alternately change direction of dip in vertical series, then the environment experienced changing directions of water flow, such as on a tidal flat having daily influx and outflow of tidal waters. Types of bedding are shown in Figure 8.1.

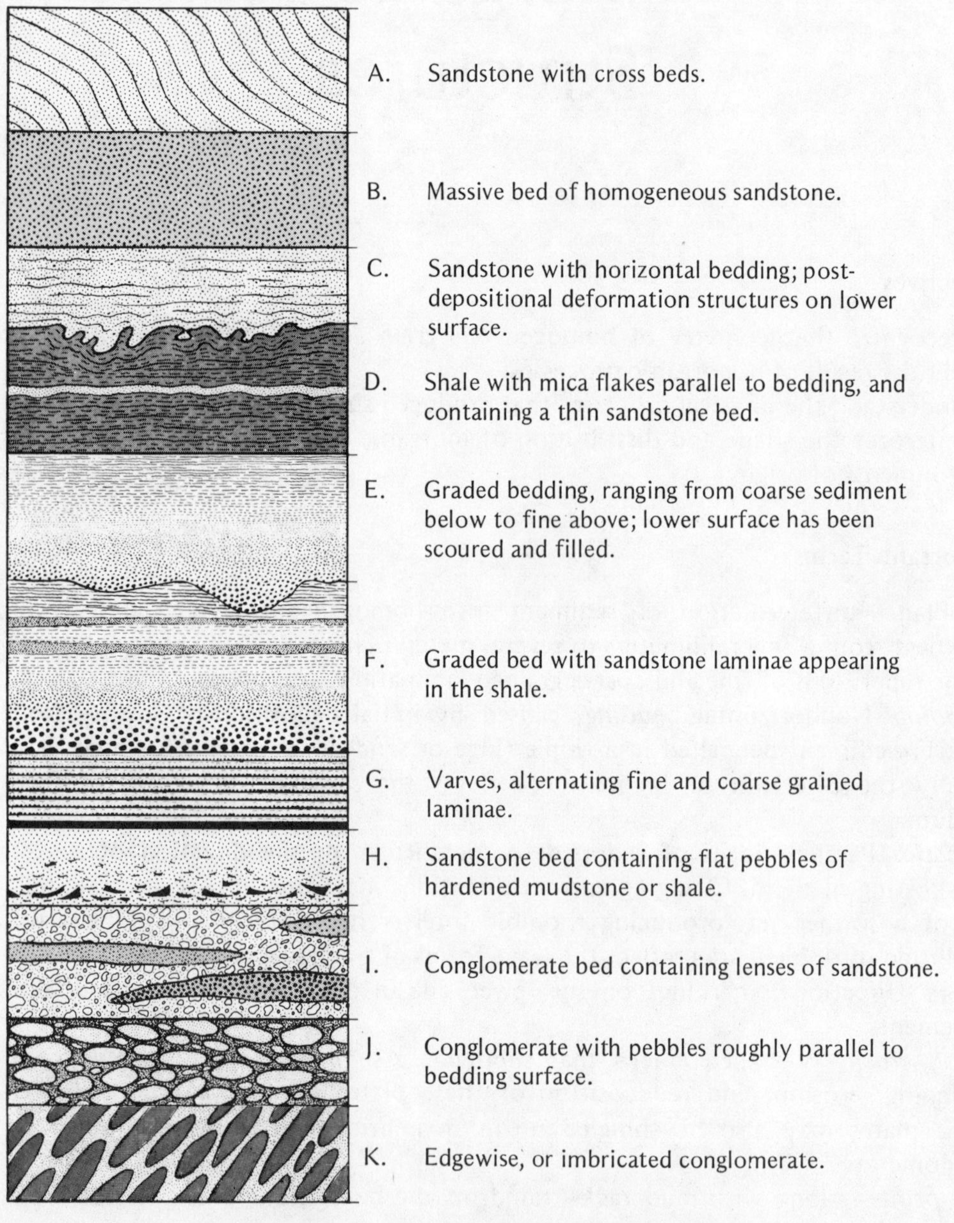

Figure 8.1 Types of beds and bedding

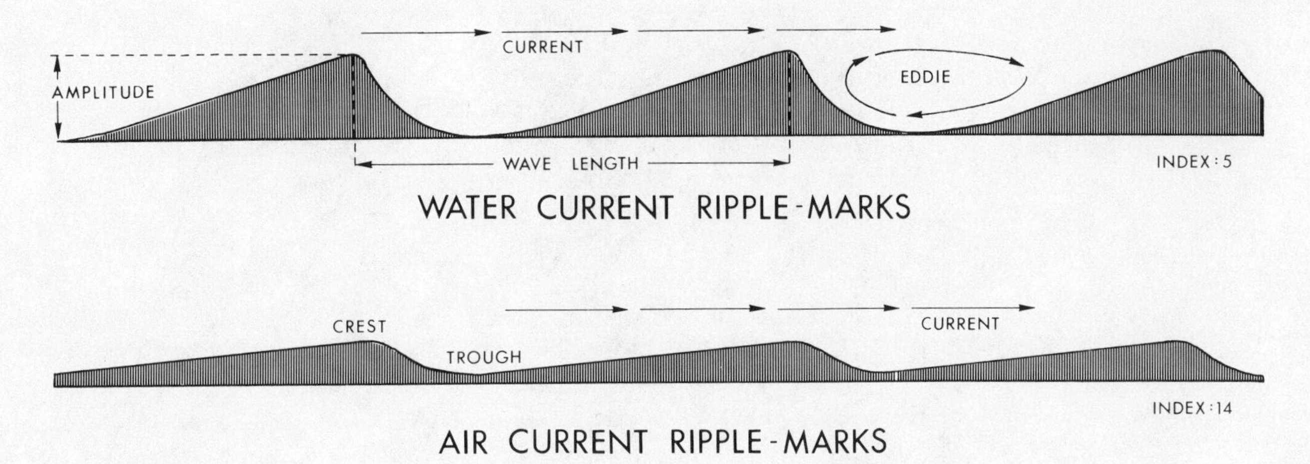

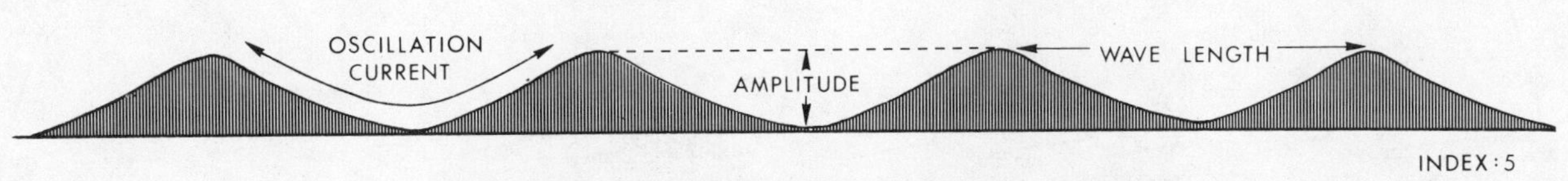

OSCILLATION CURRENT

AMPLITUDE

WAVE LENGTH

INDEX : 5

WAVE (OSCILLATION) RIPPLE-MARKS

Figure 8.2 Cross sections of ripple marks. The ripple index is the number obtained in dividing the wave length by the amplitude.

Figure 8.3 Lower surface of a bed displaying sole marks

Directional water movement also produces ripple marks (Figures 8.2, 8.4). The symmetry of ripple marks distinguishes those produced by currents from wave-produced or oscillation ripples. Oscillation ripples normally affect the substrate on a marine shelf down to about 35 meters, a level known as *wave base*. Intense wave activity caused by storms may affect even deeper portions of the marine shelf. The wave length and amplitude of the ripple marks reflect both the angle of repose of the sediment and the energy levels of the water. Sole marks are produced by scouring out of troughs in the sediment surface, which are filled in by subsequent deposition (Figure 8.3).

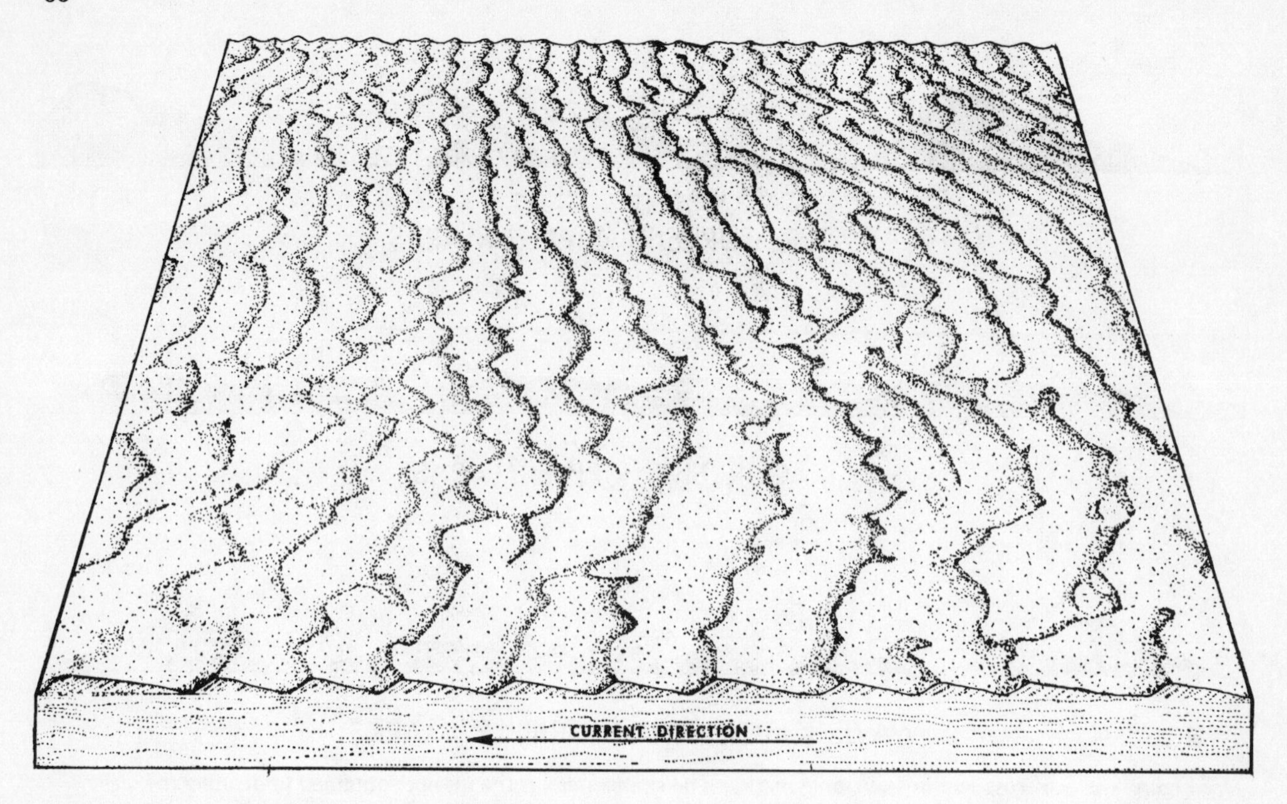

Figure 8.4 Low energy current ripples from the Chaco River, New Mexico

15 CM

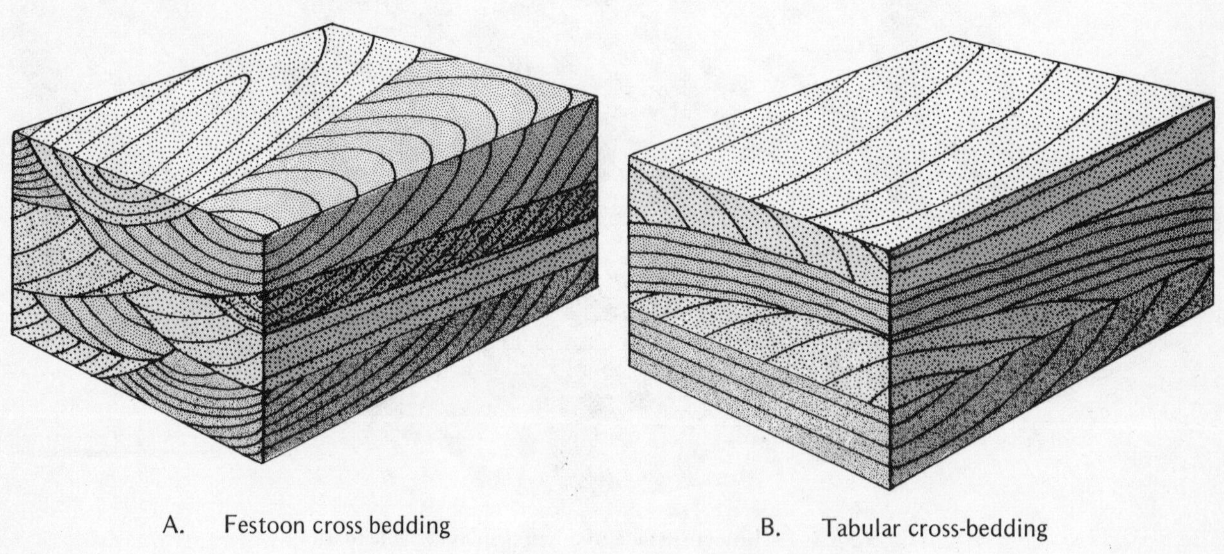

A. Festoon cross bedding B. Tabular cross-bedding

Figure 8.5 Common types of cross-bedded sets.

Cross beds will be deposited by wind or water in pulses; each pulse will leave a larger bed or unit composed of the smaller inclined beds or laminae (Figure 8.5), called *sets*, a term implying that they are composed of smaller units. The external geometry of the set depends upon the shape of the surface on which they are deposited. Sets deposited in scoured-out scoops form *festoon* cross bedding, such as in a river channel; sets deposited on flat surfaces, as on sandy beaches,

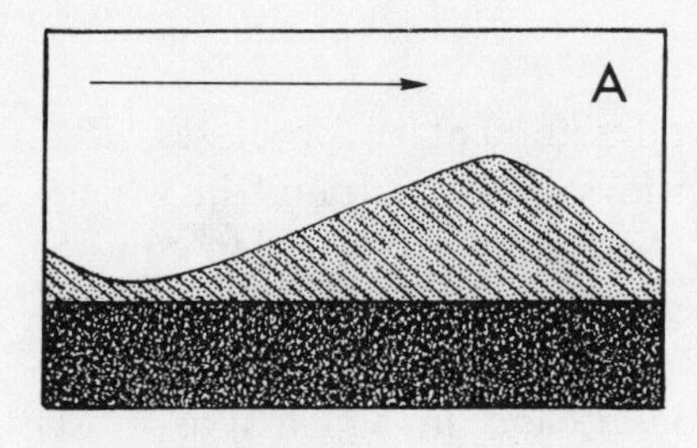

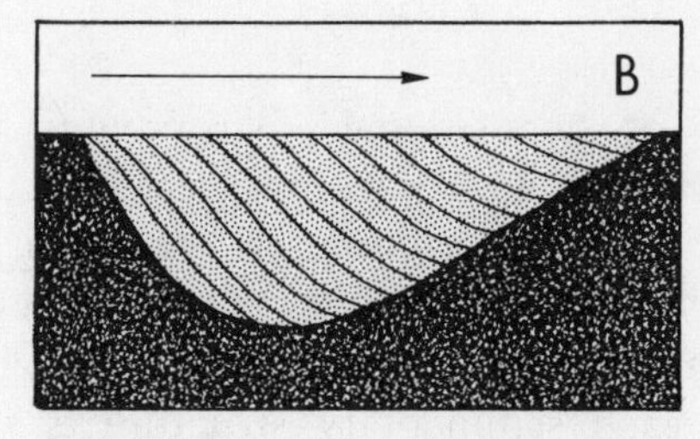

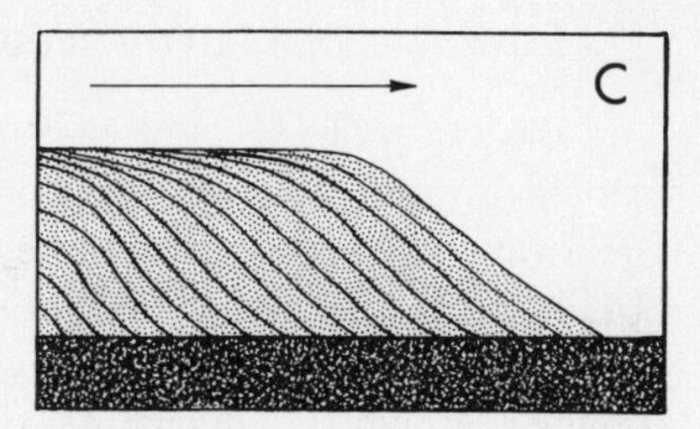

A. A small ripple ridge or large sand dune

B. A scour-and-fill structure

C. A point bar or lacustrine delta

Figure 8.6 **Sedimentary structures showing internal cross bedding**

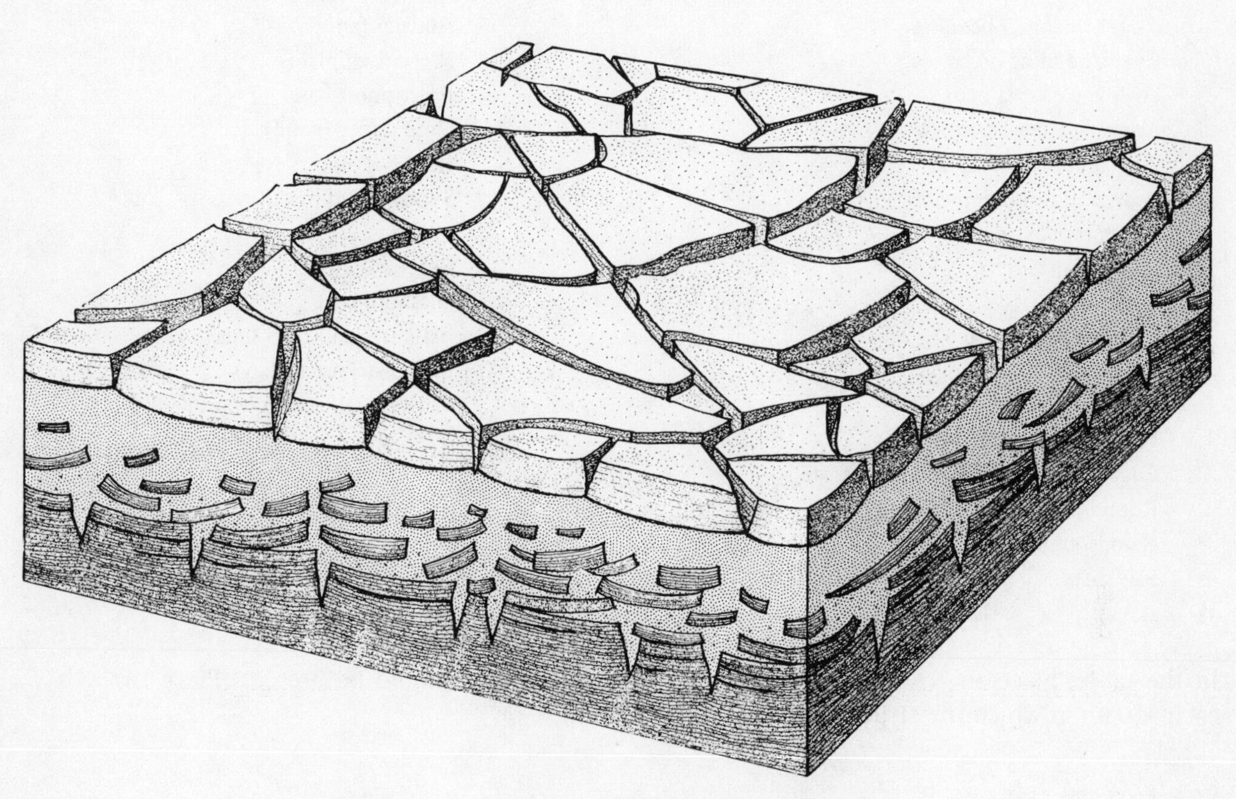

Figure 8.7 **Mudcracks and flat pebble conglomerate**

create *tabular* cross bedding. Additionally cross bed geometry depends upon the wave or wind energy and the angle of repose of the sediment.

Other primary sedimentary structures include mudcracks, flat pebble conglomerate, raindrop impressions, and salt casts. What physical processes cause the reworking of mud polygons into a flat pebble conglomerate? What do the "bomb-crater" raindrop impressions and mudcracks reveal about the depositional environment? Sometimes shrinkage cracks form polygons within a mass of sediment in colloidal layers that lose water during compaction and lithification. How could "dewatering" cracks that form within a sediment be distinguished from desiccation cracks caused by subaerial exposure to the sun? *Salt casts* are cubic cavities left within a sediment that were once occupied by salt crystals. What do they indicate about the depositional environment? Secondary structures, or those which form in unconsolidated sediments after deposition, have very little environmental meaning.

Environmental Interpretations

Based upon his knowledge of depositional environments and sedimentary processes, the student should now be able to predict which sedimentary structures should be found in which environments, and to test these predictions against real specimens. Additional information can be obtained from standard textbooks in physical geology, and the background reading given at the end of this chapter. A convenient way to summarize these predictions would be to construct a table showing in which environments each sedimentary structure is found, and in which it is most common. List the depositional environments across the top of a page, and the sedimentary structures down the left margin. Draw vertical and horizontal lines to separate the rows and columns of the table. Include the following environments and structures:

Structures

Flat, horizontal bedding
Cyclic bedding or varves
High angle cross beds
Low angle cross beds
High amplitude ripples
Low amplitude ripples
Assymetrical ripples
Oscillation ripples
Sole marks
Festoon cross bedding
Tabular cross bedding
Desiccation cracks
Flat pebble conglomerate
Edgewise (imbricate) conglomerate
Round pebble conglomerate
Raindrop imprints
Salt casts

Environments

Alluvial fans
River floodplains
River point bars
Sand dunes (wind)
Lake bottoms
Swamps and marshes
Lagoons
Beaches
Offshore bars
Estuaries and tidal flats
Interdistributary bays
Marine shelf, above wave base

In the table, place an X in each square in which the structure should be present. Place two X's in each square in which the structure should be common.

Exercise

All of the different types of sedimentary structures listed above will be set out in the laboratory, both as knowns and unknowns.

1. Identify each unknown and determine its most likely environment of origin from your table.
2. Examine the other features of each rock, including grain size, sorting, organic content, and fossils. Do not fail to use the microscope or hand lens for this work. How does this information compare with the predictions made above?
3. After having examined a representative series of real specimens, make whatever corrections might be needed in your table. Why were some of your earlier assumptions possibly incorrect?

Test Questions

1. Sediment grains and particles are frequently reworked and redeposited, often superimposing the effects of different processes, and environments on the grains. Can the same thing happen

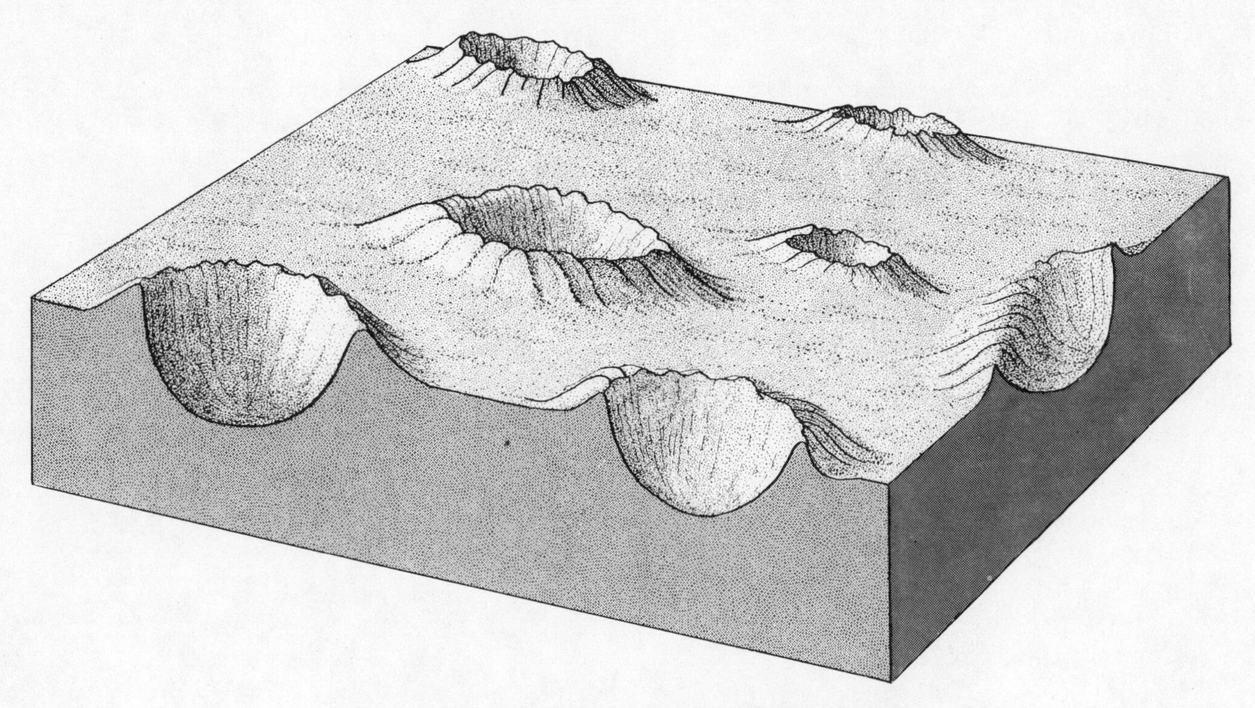

Figure 8.8 Raindrop impressions (greatly enlarged)

to a sedimentary structure? In this light, are the textural properties (size, shape, sorting) of the grains or of the larger structures more reliable in environmental reconstruction?

2. Which sedimentary structures are the most diagnostic of particular environments? Which are not particularly useful?
3. What information do some structures give that the sediment grains alone did not provide? Could field studies of sedimentary structures obtain data that could be used to reconstruct part of the paleogeography of a large region?
4. What effects will burrowing and sediment feeding organisms have on sedimentary structures? Might the presence of laminations, bedding and other structures reflect the absence of such animals? Why would these animals be present in some environments and absent in others?
5. Can all of these structures be present in carbonate as well as detrital rocks?

Background Reading

Middleton, G. V., ed., 1965, Primary Sedimentary Structures and their hydrodynamic interpretations, Soc. Economic Paleontologists and Mineralogists, Spec. Pub. 12.

Pettijohn, F. J., and Potter, P. E., 1964, Atlas and Glossary of Primary Sedimentary Structures, Springer-Verlag, New York.

Selley, R. C., 1970. Ancient Sedimentary Environments, Cornell Univ. Press, Ithaca, New York.

Shrock, R. R., 1948, Sequence in Layered Rocks, McGraw-Hill Book Co., New York, p. 63-326.

Chapter Nine

Paleontology
Trace Fossils

Objectives

1. To examine the disruption of physical sedimentation by the activity of organisms.
2. To determine the environmental meaning of sedimentary structures produced by organic activity.
3. To examine the meaning of assemblages, or communities, of trace fossils.

Important Terms

Trace fossil—A sedimentary structure produced in unconsolidated sediment by the activities of an organism (borings into hard substrata are only rarely preserved). Trace fossils include tracks, trackways, trails, shafts, and tunnels.

Bioturbation structure—Any disruption of sedimentation by organic activity, including trace fossils lacking distinctive geometry, caused by intensive reworking of the sediment, sometimes producing a burrow-mottled texture.

Tracks—Isolated impressions left on the substrate surface by individual foot structures (podia).

Trackway—A continuous series of tracks made by the continuous, linear movement of a single animal.

Trail—A continuous groove or furrow left on the substrate surface by an animal that did not leave separate footprints, but plowed through the sediment with part or all of its body.

Shaft—A vertical cylindrical burrow, or the vertical part of a complex burrow system.

Tunnel—A horizontal cylindrical burrow, produced *within* the sediment, or the horizontal part of a complex burrow system.

Spreiten—(singular, *spreite*) Bladelike or spiral series of many closely spaced parallel or concentric burrows, produced by the migration of a single animal's burrow upwards, downwards, or sideways through the sediment.

Discussion

The same animal may produce many different types of trace fossils, depending upon its behavioral activities. Figure 9.1 shows crawling, foraging, feeding, and resting traces, all attributed

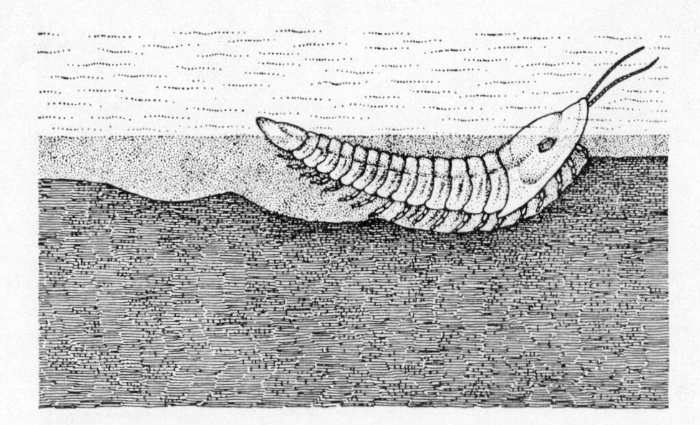

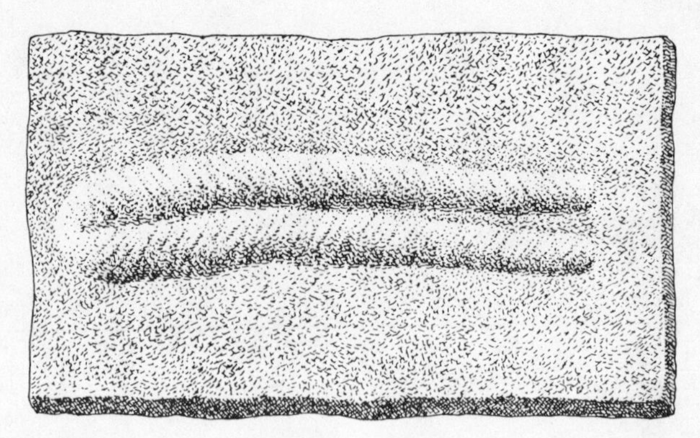

A. Crawling tracks

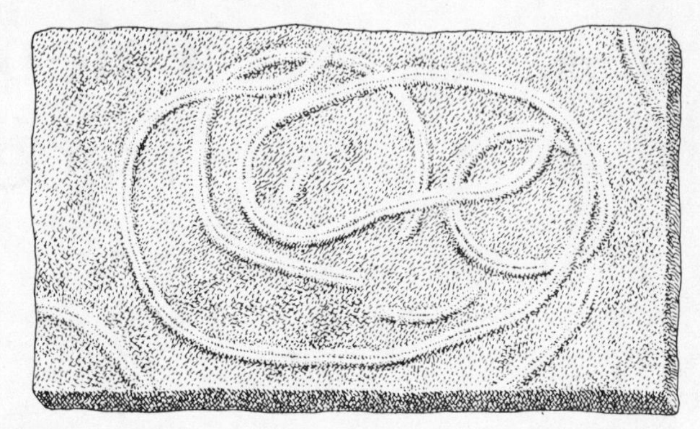

B. Foraging tracks

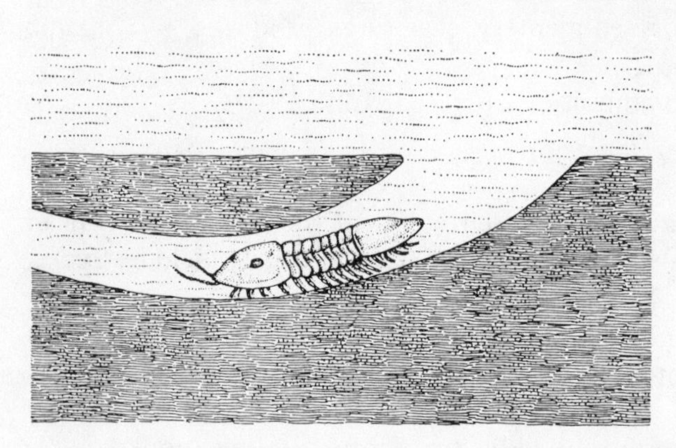

C. Feeding burrow

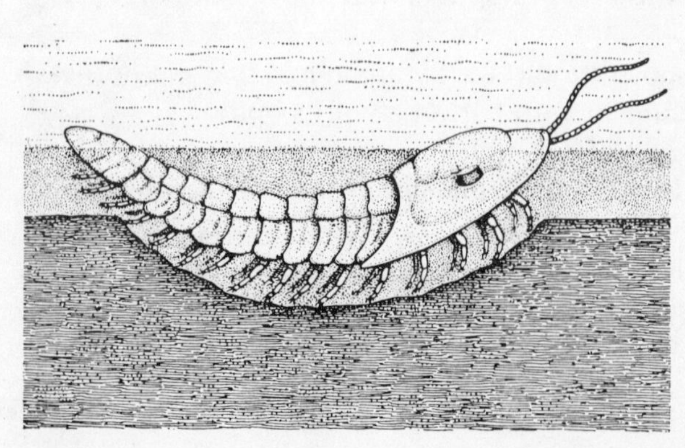

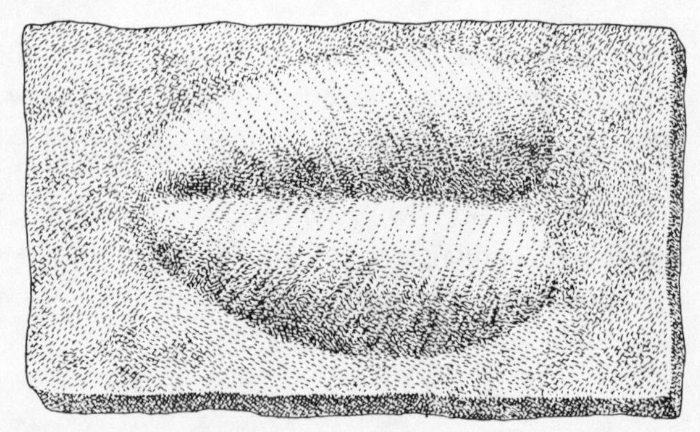

D. Resting tracks

Figure 9.1 **Different types of trails and burrows made by a single trilobite.** From "Fossil Behavior," Adolf Seilacher.

to the same type of trilobite. Conversely, many different animals may produce the same type of trace fossil, because of similar behavior. A U-shaped burrow (Figure 9.2A) may be formed as a dwelling burrow by a great variety of suspension feeders. Burrows made by the same organism may have different shapes and densities in different environments (Figure 9.2D). Thus it is usually impossible to determine which animal has made a trace, and trace fossils of differing morphology should not be classified with Latin names, as other fossils are. Rather they would be better described by simple geometric terms that describe the ancient behavior.

Tracks and trackways are common fossils, and have been produced by organisms ranging from insects to dinosaurs, birds, and fossil men. Tracks reflect simple locomotion, and do not reveal much about behavior. Trails and burrows are produced mainly by invertebrate animals. Crawling trails (Figure 9.1A) show only locomotion, but foraging trails (Figure 9.1B) provide evidence of feeding behavior. Burrows may result from sediment feeding (Figure 9.1C), resting (Figure 9.1D), or dwelling (Figure 9.2A). Burrows may be horizontal tunnels (Figure 9.2B), or vertical shafts (Figure 9.2C).

Environmental Models

Trace fossil assemblages have been observed to fall into five main groups (Figure 9.3); these groupings probably reflect community structures similar to those studied in Chapter 6. The "*Skolithos*," "*Glossifungites*," and "*Cruziana*" communities should correspond in a general way to the three detrital communities of skeletal fossils. The "*Zoophycus*" and "*Nereites*" communities presumably represent deeper water than the continental shelf, representing bathyal and abyssal habitats (Figure 3.7). The idea that these trace fossils represent such abnormally great depths should be considered a model rather than a firm conclusion.

Spreiten-type burrows may reflect either the animal's response to sedimentation and growth, or programs of simple or highly efficient sediment mining. In shallow water sands, the upward migration of the burrow may be the simple result of high sedimentation rates. In areas where sedimentation is slow, the burrow may be migrating downward to accommodate the growing animal as it becomes larger. Vertical burrows are most commonly made by filter feeders, to serve as dwelling burrows. More filter feeders are found in certain habitats because more food is suspended in the water. Also the deep, vertical burrows afford protection from the greater number of predators in zones where suspended food is abundant. Most of the burrows around a modern barrier island (Figure 9.4) are shafts, although different shallow water environments show small changes in shaft geometry, mostly due to burrowing by different types of animals.

In offshore, deeper water habitats, less food is suspended in the water, so more animals have adapted to mining food out of the sediment. The percentage of filter to sediment feeders among the trace fossils reflects the feeding regime of the community. If less food is present in the sediment, then sediment mining must be carried out more efficiently, as in the "*Nereites*" community. Such efficient sediment mining may also reflect the absence of predators from habitats where the supply of suspended food is low. Certainly extensive surface sediment grazing causes an animal to be exposed almost constantly to predators.

The abundance of recognizable burrows unfortunately does not reflect the abundance of the burrowers. In nearshore environments in which burrowers are profuse, the sediment may be completely homogenized by organic reworking, and would lithify to a massive, featureless rock. The same sediment may have passed through the guts of several different organisms, which may greatly influence its grain size and sorting. Burrows are best preserved when the burrowers are not especially numerous.

Evolution

A traditional idea in historical geology is that the "sudden" appearance of animal skeletons in earliest Cambrian strata reflects only their simultaneous acquisition of hard parts, not the explosive

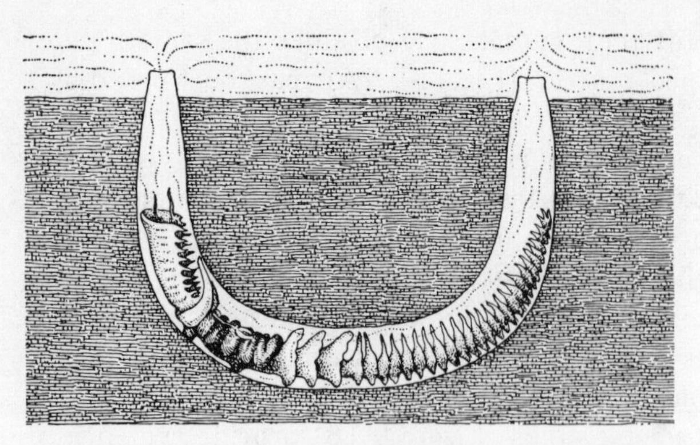

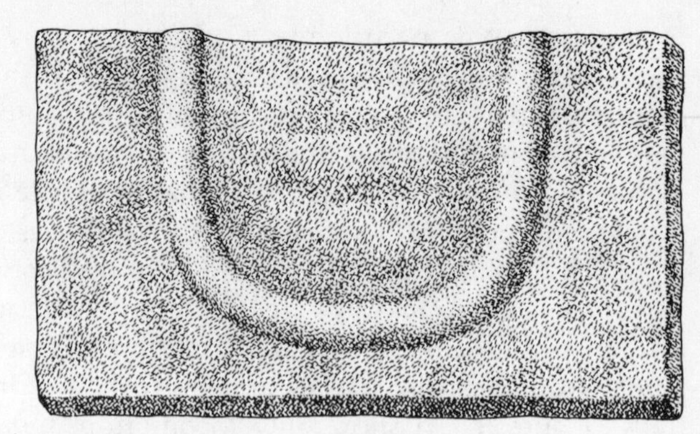

A. Dwelling burrow of a marine annelid. From "Fossil Behavior," Adolf Seilacher.

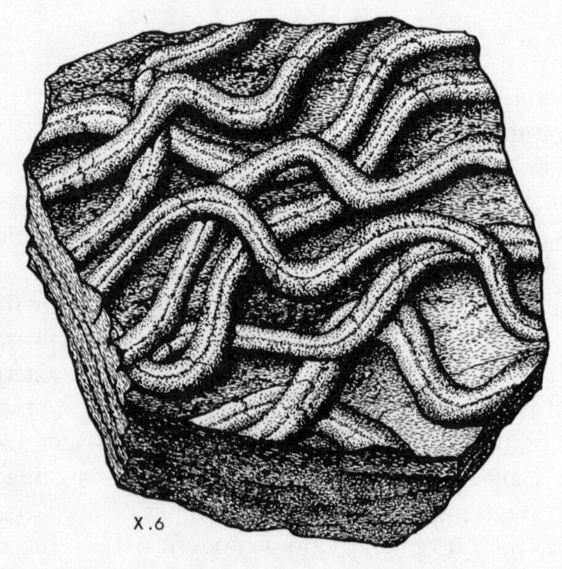

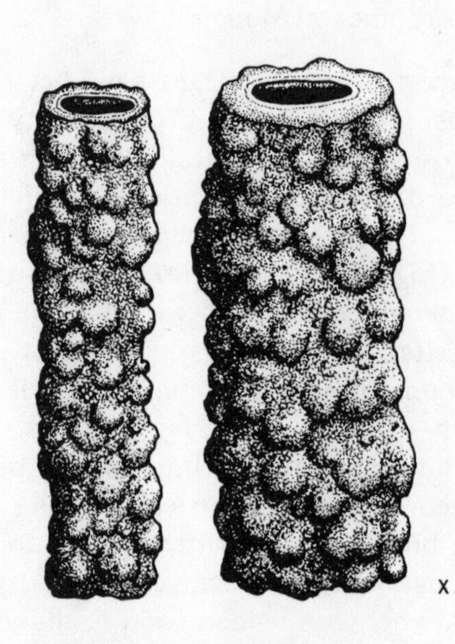

B. Slab of Ordovician limestone with the tunnels made by a wormlike animal.

C. Sections of lithified *Callianassa* shafts from the Miami Ooolite, Florida Keys.

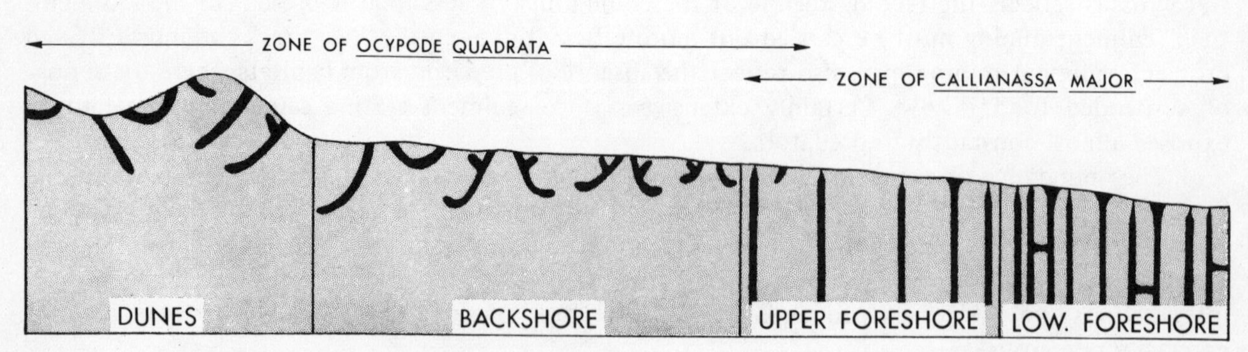

D. Shafts made by the ghost crab, *Ocypode,* and the ghost shrimp. *Callianassa,* in the beach-dune complex of Sapelo Island, Georgia.

Figure 9.2 Burrows and tracks

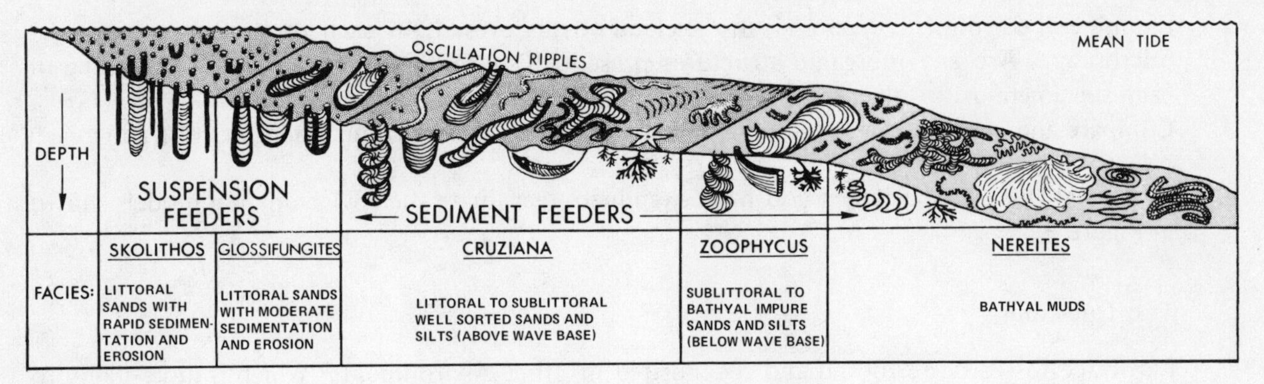

Figure 9.3 Trace fossil communities

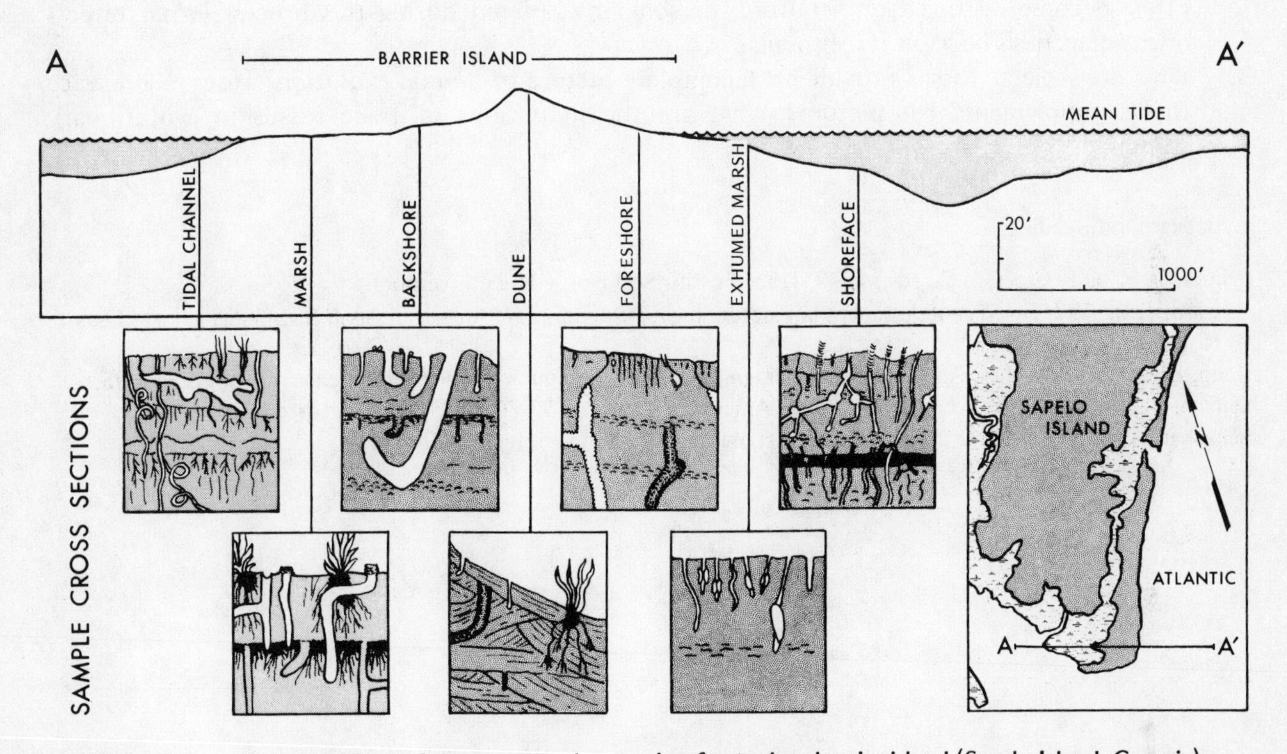

Figure 9.4 Variations in shaft burrows on the margin of a modern barrier island (Sapelo Island, Georgia)

diversification of animal life. If this idea is true, then Precambrian rocks should contain diverse assemblages of trace fossils, reflecting the activities of the unshelled animals. Unfortunately for this concept, trace fossils show an "explosion" in diversity and complexity at the base of the Cambrian similar to that of the shelled organisms.

Trace fossils also show progressive evolutionary trends in animal foraging behavior through geologic time. As with the skeletal fossils, the morphology of trace fossils has evolved from the simple in the direction of increased complexity.

Exercise

Examples of different types of trace fossils will be set out as known and unknown specimens.

1. Identify each unknown. Determine to which community it belongs, and describe its most likely environment of origin.

2. Examine the sediment type and any skeletal fossils present in each rock, using a binocular microscope. Are any inorganic structures present? Determine the environmental meaning of each specimen using this information.
3. Compare and contrast the two sources of information. Do they complement or contradict each other?
4. In the light of the specimens you have examined, reevaluate the environmental models shown in Figure 9.3

Test Questions

1. Can trace fossils be reworked and redeposited in other environments? Which is more likely to be exactly in its original habitat, a trace fossil or a skeletal fossil?
2. *Diagenesis* is the process whereby a sediment becomes lithified. Shells and other skeletal fossils often become dissolved, mineralized, or somehow altered during this process. What effects might diagenesis have on trace fossils?
3. Why do skeletal fossils provide an incomplete picture of animal evolution? How might trace fossils supplement that picture? What are the limitations of trace fossils in evolutionary studies?

Background Reading

Crimes, T. P., and Harper, J. C., eds., 1970, Trace Fossils, Seel House Press, Liverpool.

Howard, J. D., 1971, *in* Recent Advances in Paleoecology and Ichnology, American Geol. Inst. Short Course Lecture Notes, p. 148-268.

Raup, D. M., and Seilacher, A., 1969, Fossil foraging behavior: computer simulation: Science, v. 166, p. 994-95.

Seilacher, A., 1967, Fossil behavior: Scientific American, v. 217, p. 72-80.

Seilacher, A., 1967, Bathymetry of trace fossils: Marine Geology, v. 5, p. 189-200.

Chapter Ten

Paleontology
Algal Stromatolites

Objectives

1. To study the origin and morphology of different types of algal stromatolites.
2. To understand their environmental meaning, using their mode of origin as well as comparisons with living stromatolites.

Important Terms

Stromatolite—Laminated sedimentary structure formed by blue-green algae or, in some cases, bacteria; structures having the overall form of sheets, pillars, or spheres.

Blue-green algae—Procaryotic unicellular organisms (Chapter 3), rarely preserved as fossils, but responsible for forming stromatolites.

Algal mat—Dense, sheetlike mass of intertwined algal filaments; may contain green algae and bacteria as well as blue-greens.

Photosynthesis—The biochemical process whereby sunlight is used to produce food; the reverse process, whereby food is consumed to produce energy, is *respiration*.

Discussion

Stromatolites are very thinly laminated three-dimensional bodies of lithified lime mud (micrite). They are *neither* skeletal fossils nor trace fossils, but are sedimentary structures produced by a combination of organic and inorganic processes. They are mostly formed by the deposition or chemical precipitation of lime mud on top of, within, or underneath an algal mat. Algal mats are thick, sheetlike masses of intertwined algal filaments. A single algal mat may contain numerous different genera and species of living algae, up to twenty-eight species having been recorded in a single mat in southern Florida. Also, as will be discussed below, the same algal mat may produce morphologically different structures in different environments. Thus, as for trace fossils, there is no value in assigning Latin names to stromatolites of different form. Rather it is more useful to describe them by simple geometric formulas that reflect their conditions of origin.

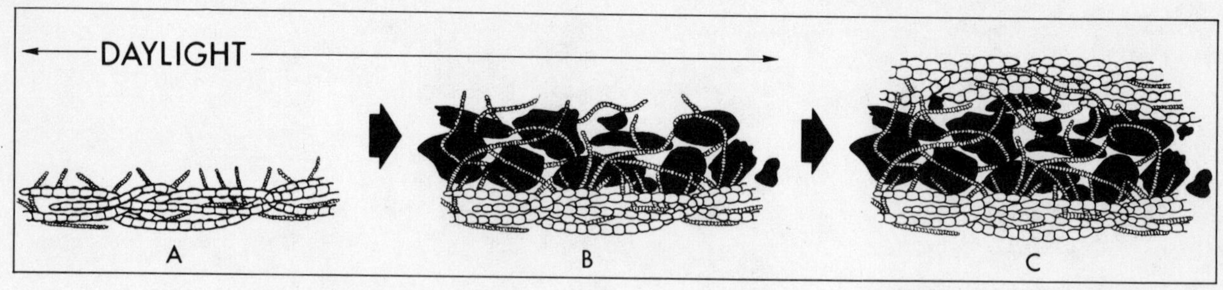

Figure 10.1 Diagram illustrating the entrapment of sediment particles by an algal mat; dark masses are sediment grains.
A. Uncovered mat at beginning of daylight period.
B. Sediment trapping during daylight.
C. Regrowth of mat and sediment binding during darkness.

These structures originate from the simple entrapment of sediment on the mucilaginous sheaths of the algal filaments (Figure 10.1). The microstructure of the algal filaments may be observed in rare, well-preserved specimens. Each layer of sediment forms a single lamina. The laminae may be flat lying, gently undulating, strongly arched, updomed, or columnar in mats covering the bottom, or spheroidal in mats coating loose objects that roll freely around on top of the bottom. The form of the laminae reflects the form of the mat on which the sediment was caught.

Origin of the Lime Mud

Algal stromatolites rarely contain any detrital mud or sand, and a detrital origin of the lime mud seems unlikely. It seems more probable that the lime mud originates in the immediate vicinity of the algal mat by chemical precipitation that is controlled and mediated by organic processes. At least two models are possible:

A. Photosynthesis

The photoperiod (time of continuous daylight) controls the duration and intensity of photosynthesis, in which the following reaction takes place:

$$CO_2 + H_2O + \text{Light} \longrightarrow C_n(H_2O)_n \text{ (sugar)} + O_2$$

At night, this equation is reversed, and respiration takes place:

$$C_n(H_2O)_n \text{ (sugar)} + O_2 \longrightarrow CO_2 + H_2O + \text{energy}$$

Many geologists believe that in the Precambrian, the burial of unoxidized organic material resulted in the liberation of free oxygen into the atmosphere, because it was not used up either in respiration or in the complete oxidation of the organics. Thus algal mats may have played some part in the formation of an oxygenated atmosphere.

During daylight, according to the above reactions, CO_2 is removed from the water overlying and within the mat, but during the night CO_2 builds up in the water. Because dissolved CO_2 is a weak acid, calcium carbonate is more soluble in the water during night than day. During daylight the water becomes more alkaline, and solid $CaCO_3$ can be precipitated as mud-sized particles, both within the algal mat and on its upper surface, where it is entrapped by the sticky thalli.

B. Anaerobic Bacteria

Reducing conditions in the lower layers of the mat resulting from bacterial decay could produce ammonia and precipitate $CaCO_3$:

$$NH_3 + H_2O \longrightarrow NH_4OH$$

$$2NH_4OH + Ca(HCO_3)_2 \quad \text{(dissolved bicarbonate)} \longrightarrow (NH_4)_2CO_3 + 2H_2O + CaCO_3 \quad \text{(solid)}$$

This process could account for observed gradations between pure mat and pure "algal" micrite by the bacterial destruction of the mat and replacement by $CaCO_3$. Discrete loci of initial micrite precipitation may be related to more favorable areas for bacterial colonization. Updoming of stromatolites may be produced by either substrate irregularities or gas formation related to bacterial decay.

Bacteria alone may form stromatolites, and structures resembling the Precambrian "*Conophyton*" stromatolites are forming today by bacterial precipitation of carbonate from the waters of hot springs. Oolites (Chapter 7) might simply be spheroidal bacterial stromatolites; evidence includes dark organic matter between the laminae as well as assymmetrical oolites forming in quiet water. Other chemical sediments in which bacteria play a large, if poorly understood, role are chert (amorphous SiO_2), pyrite, and possibly chalcocite.

Environmental Significance

Three morphological types are known, as well as compound or transitional forms (Figure 10.2):

LLH—Laterally linked hemispheroids with continuous laminae; domes close (LLH-C) or widely spaced (LLH-S), separated by flat areas.

SH—Stacked hemispheroids forming columns or clublike heads separated by nonlaminated sediment; domes with a constant diameter (SH-C) or a variable diameter (SH-V).

SS—Spheroidal structures surrounding a nucleus that rolled freely on the bottom; laminations concentric (SS-C), randomly discontinuous (SS-R), or inverted (SS-I).

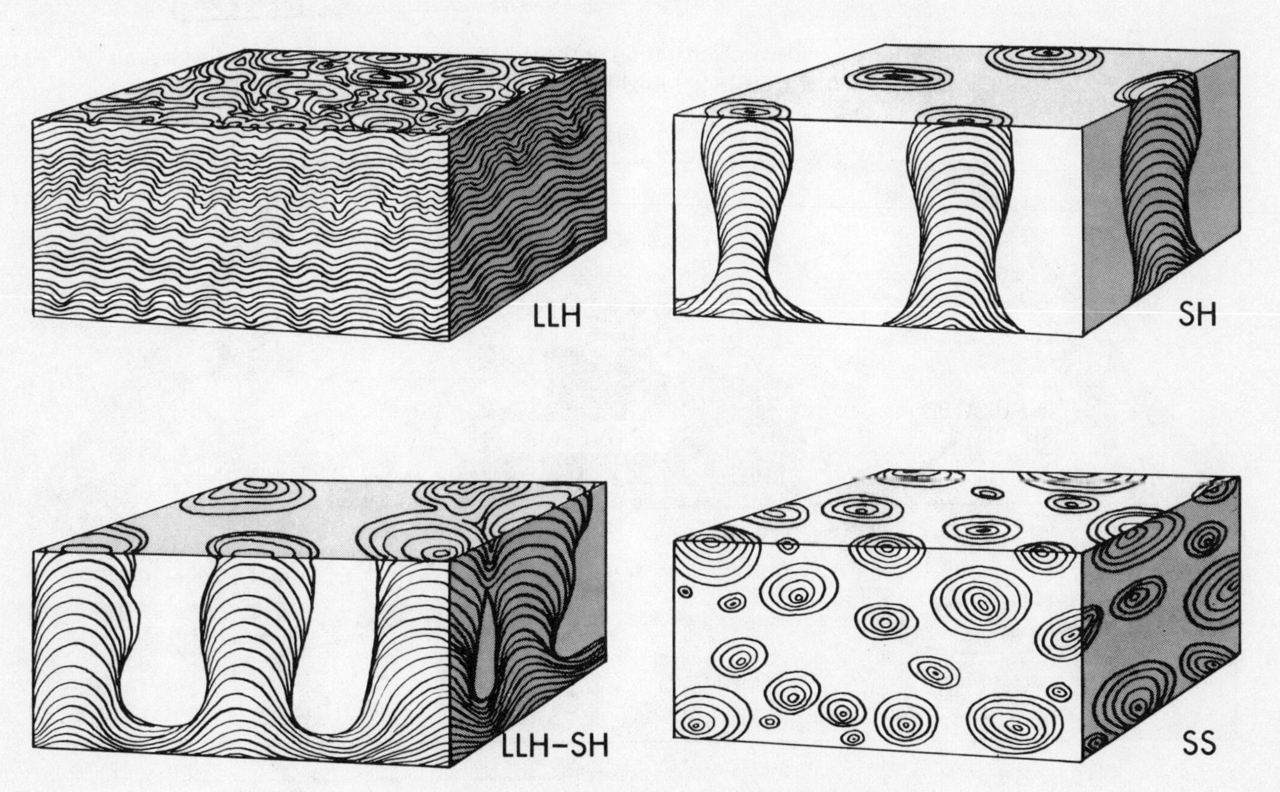

Figure 10.2 Morphology of algal stromatolites

The distribution of these types in a modern tidal area is shown in Figure 10.3. Types LLH and SH are usually intertidal, and require periodic desiccation for their development. LLH forms mostly on mudflats in the protected locations of reentrant bays and behind barrier islands and ridges where wave action is usually slight. SH is formed in areas of more vigorous wave action, commonly on exposed intertidal headlands, principally by initial relief on the substrate, induration, periodic wetting and desiccation, and scouring or heavy sedimentation in the interarea between the clublike heads (Figure 10.4). Early lithification, speeded up by the drying process, prevents collapse of the

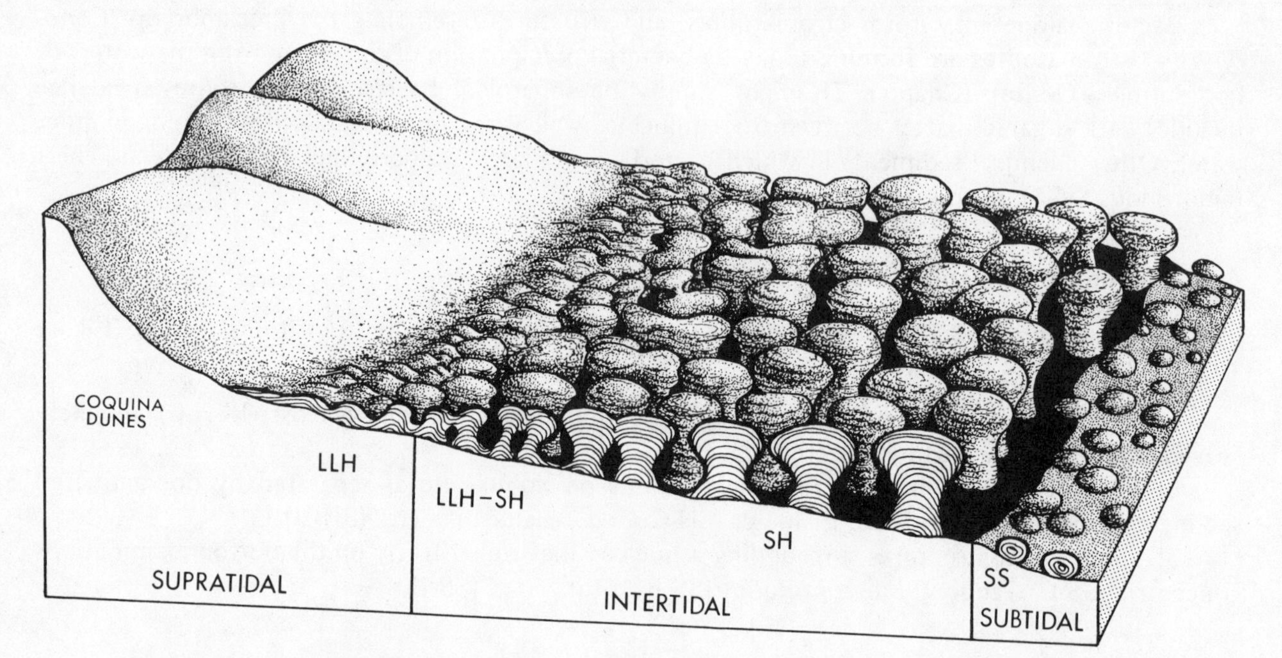

Figure 10.3 Distribution of modern stromatolites in the tidal area of a modern hypersaline lagoon (Shark Bay, Western Australia); highly generalized.

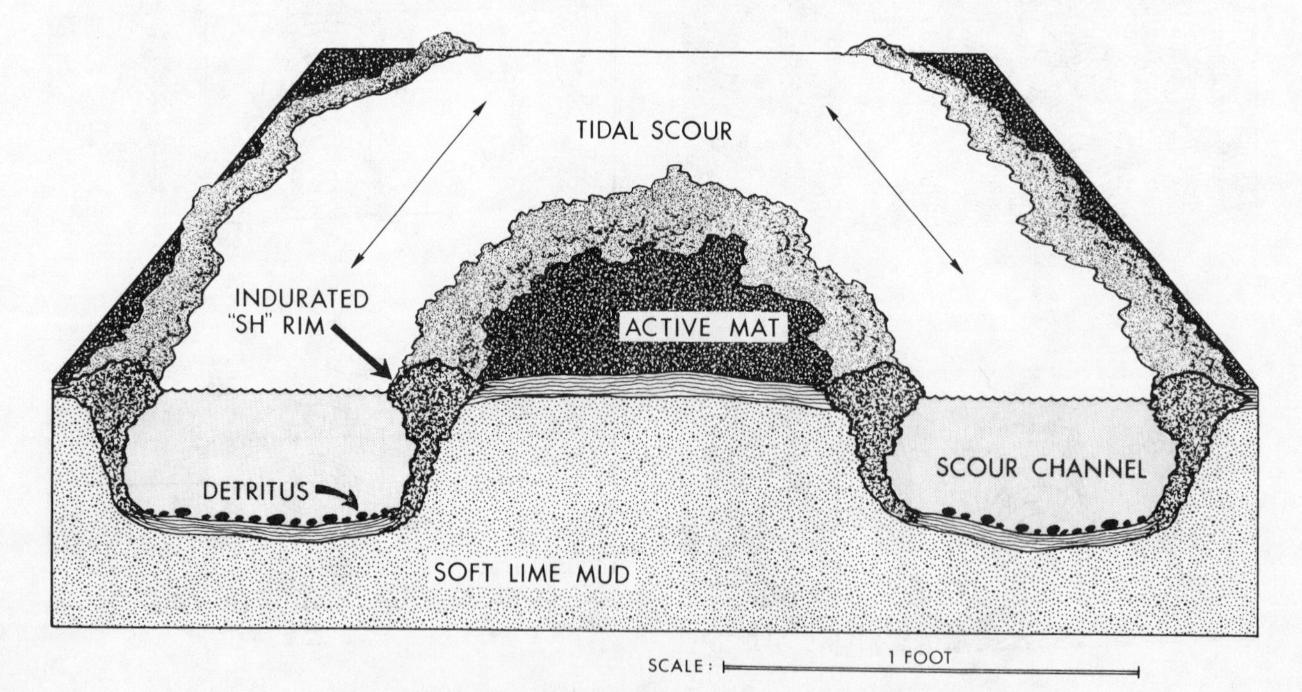

Figure 10.4 Formation of clublike heads (SH) by tidal scour and rapid induration of the rim

structures. Modern algal stromatolites are forming only in tropical and subtropical regions where aragonite precipitated from sea water aids in the induration (hardening) of these structures. Type SS is found only in permanently submerged shoal water areas, or areas low in the intertidal zone, where the algal-coated bodies are subject to more or less continuous rolling.

Algal stromatolites can also form along the margins of hypersaline lakes, as along the shoreline of Great Salt Lake, Utah. Stromatolites have been reported from salt lakes in western Australia, and from Tertiary lakes in the Rocky Mountain region. Recent interpretations of the Belt stromatolites (Late Precambrian, Montana) suggest an origin in hypersaline lakes formed in the center of an ancient fault-bounded valley, much like the present Dead Sea. One of the Belt boundary faults is well known; to the north of it there are thick alluvial fan deposits (the LaHood facies), which grade rapidly into fluvial clays and point bar sands, which in turn grade into the stromatolite-bearing carbonates of the lacustrine shoreline.

Because stromatolites are characteristic of tidal flats and hypersaline lagoons, they often display desiccation cracks and are frequently dolomitized. Compound structures reflect conditions that changed during the growth of the stromatolites.

Exercise

1. Identify all of the unknown stromatolites, by comparison to known specimens and to Figures 10.2 and 10.3.
2. Identify the sediment type, and any associated skeletal fossils, trace fossils, or inorganic sedimentary structure.
3. Synthesize all available information to determine the most probable conditions of origin of each specimen.

Test Questions

1. Is it correct to refer to stromatolites as "fossil algae"? Why or why not?
2. In what areas of the United States today might stromatolites be forming? Could stromatolites form along detrital-sediment shorelines?
3. What characteristics of modern fault-bounded salt lake basins could be used to test the idea that the Belt sediments were deposited in such an environment? Consider sediment thickness, extrusion of igneous flows, and sedimentary structures, including mudcracks, flat pebble conglomerates, and salt casts.
4. Why are stromatolites useful in paleogeographic mapping?

Background Reading

Logan, B. W., 1961, *Cryptozoon* and associate stromatolites from the Recent, Shark Bay, Western Australia: Jour. Geology, v. 69, p. 517-33.

Logan, B. W., Rezak, R., and Ginsburg, R. N., 1964, Classification and environmental significance of algal stromatolites: Jour. Geology, v. 72, p. 68-83.

Chapter Eleven

Sedimentary Petrology
Origin of Dolomite

Objectives

1. To determine, as precisely as possible, the succession of depositional environments in a vertical sequence of dolomitized sedimentary rocks.
2. To determine the lateral distribution of the ancient environments from the vertical sequence, using a principle of stratigraphy known as Walther's Law.
3. To determine the environment in which the dolomite originated, and which model for the origin of dolomite best fits the rock sequence.

Important Terms

Walther's Law—The idea that the vertical succession of rock types observed at a single locality reflects the lateral succession of environments that have migrated over that depositional site.

Sabkha—Coastal plain underlain by porous sediments; surface evaporation causes landward flowage of seawater through the sediment.

Seepage refluxion—Seaward flowage of brines concentrated by evaporation through porous sediment.

The Sabkha Model

Two areas where dolomite is forming today are the Abu Dhabi sabkha in the Arabian Gulf (Figure 11.1) and Andros Island in the Bahamas (Figure 7.7, 11.2). In both areas seawater of normal or higher than normal salinity permeates carbonate sediments originally composed of aragonite. These are pelletal lime muds in the Bahamas and algal laminated lime muds (stromatolites) in the Arabian Gulf. Evaporation in the supratidal zone causes flowage of sea water upward to the evaporative surface. The dolomite originates by the emplacement of the magnesium ion from seawater into the carbonate lattice, altering it to $CaMg(CO_3)_2$. This replacement starts as the growth of microscopic dolomite rhombs around small nuclei and may proceed to the complete dolomitization of the sediment, involving complete recrystallization and obliteration of primary

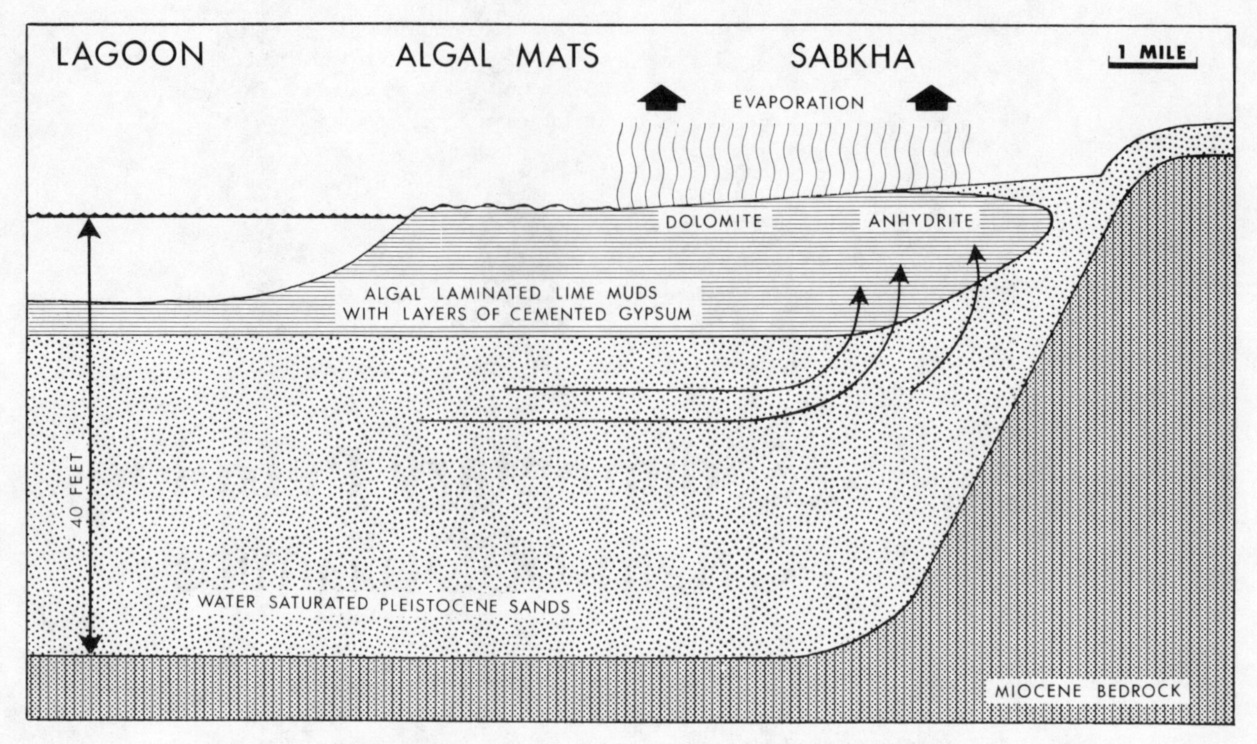

Figure 11.1 The Abu Dhabi sabkha in the Arabian Gulf

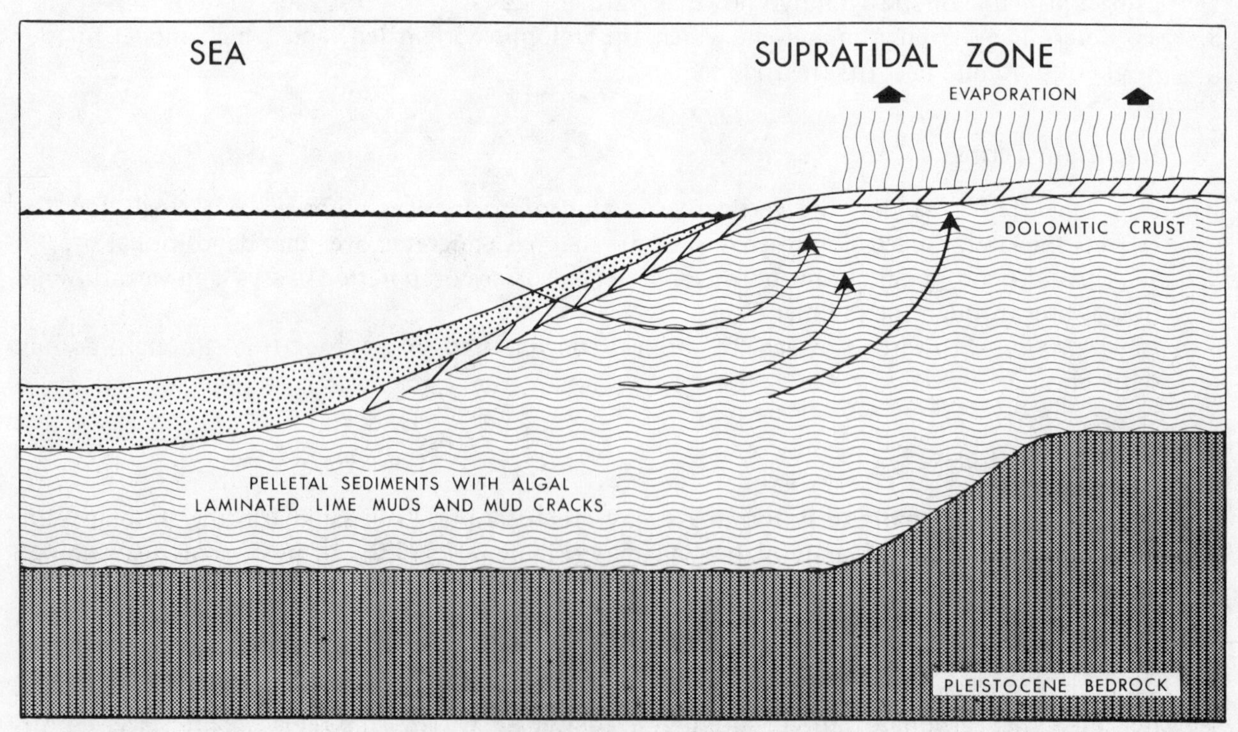

Figure 11.2 The dolomitic crust on Andros Island, Bahama Bank; seaward the crust is older, having been buried as a result of the rise of sea level; thickness of crust exaggerated, actually not thicker than 3 cm.

textures, including fossils. Such rocks often have sedimentary structures characteristic of the supratidal zone, including LLH stromatolites and mud cracks (Figure 11.5).

The Seepage Refluxion Model

Alternatively, hypersaline waters entrapped in a coastal lake or barred lagoon could possibly flow through porous carbonate sediment seaward (opposite to the flowage direction in the sabkha), perhaps by osmotic pressure (Figures 11.3, 11.4). Again, flowage through aragonitic sediments can cause recrystallization to dolomite. The major difference in the two models is that the seepage model has only a limited amount of brine to pump through the sediment, whereas the amount of sabkha brine is unlimited. Consequently sabkha dolomitization could be more complete or intense than that of seepage refluxion. Secondly, the sabkha sediments affected are supratidal; the seepage mechanism may affect intertidal and subtidal sediments. Seepage refluxion may result in only limited dolomitization of the more porous structural units of a carbonate sediment. For example, only the coarser carbonate sand infilling burrows may be dolomitized, and the lime mud matrix not be altered, producing a burrow-mottled limestone-dolomite texture. Unfortunately, seepage refluxion has not been observed in any modern environments, and may have value only as a model.

Exercise

A collection of samples representing the vertical succession of rock types at a single locality will be set out in the lab. The rock samples containing dolomite will be identified and so labeled by your instructor. Simple acid tests are not reliable; thin section work, staining, and other methods are more conclusive.

1. Describe the sediment type, fossils, and sedimentary structures in each sample. Determine the depositional environment of each sample as precisely as possible, using all available information.

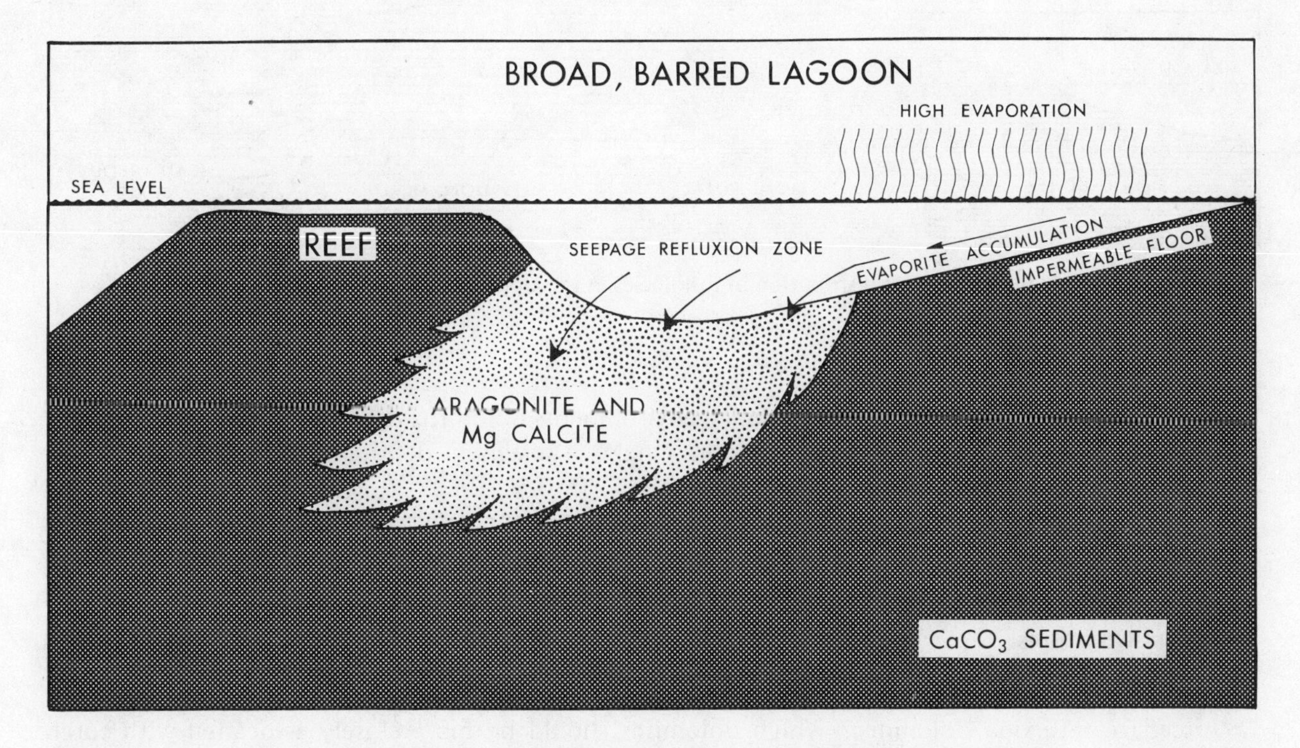

Figure 11.3 Seepage refluxion from an evaporitic lagoon

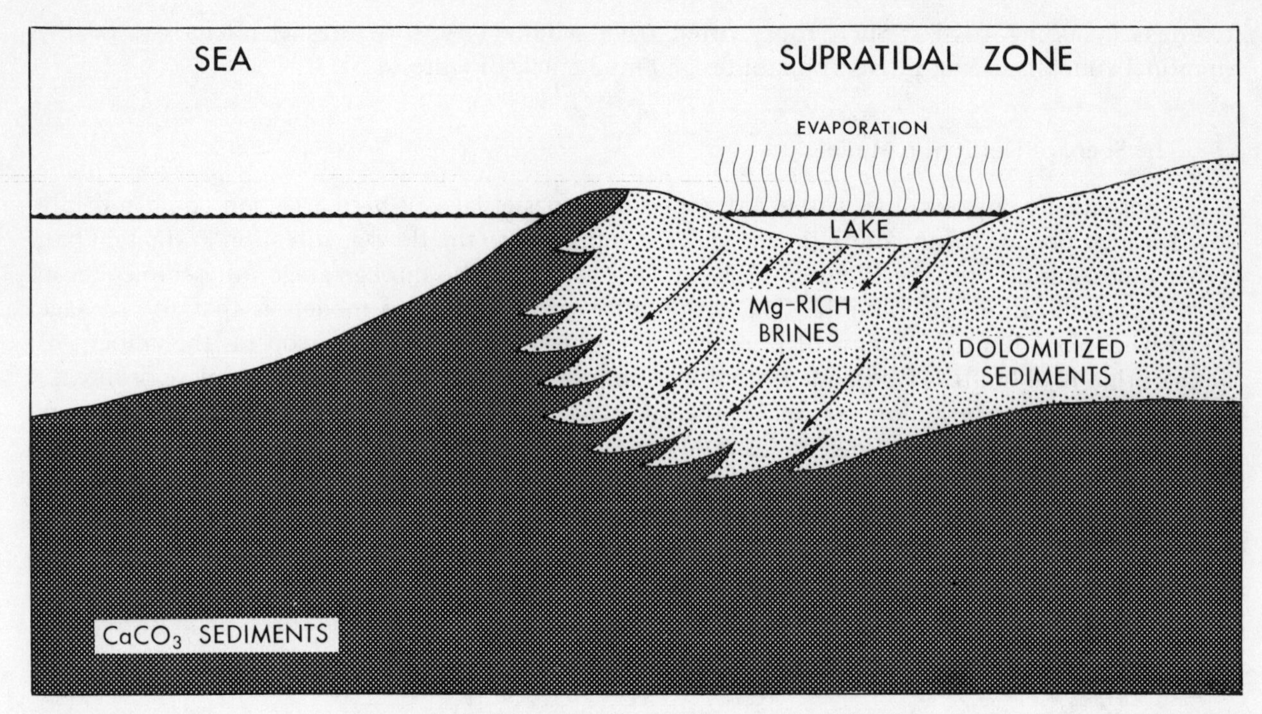

Figure 11.4 Seepage refluxion from a supratidal lake

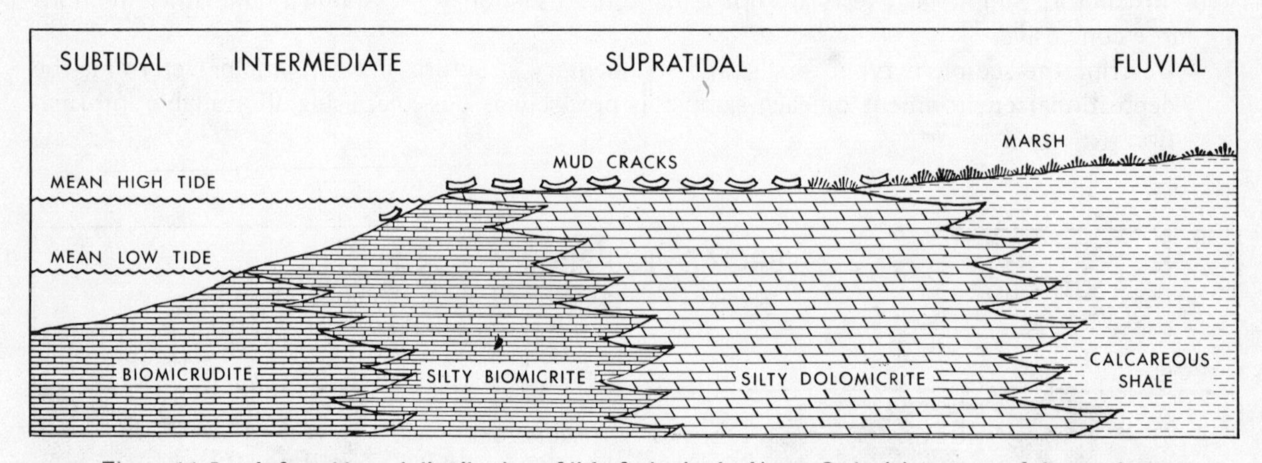

Figure 11.5 Inferred lateral distribution of lithofacies in the Upper Ordovician strata of the southern Appalachians.

2. Using Walther's Law, determine in which environments dolomitization took place, and to what degree. Using Figure 11.5 as a model, make a diagram of the lateral distribution of environments.
3. Which of the two models (sabkha or seepage refluxion) appears to provide a better explanation for the origin of the dolomites studied above?

Test Questions

1. What sedimentary structures and fossils should be associated with sabkha sediments? With seepage refluxion dolomites? Which dolomites should be more closely associated with coral reefs?

2. What climatic conditions are necessary for the development of dolomite? Is dolomite a paleoclimatic indicator?
3. Using the sabkha model, under what conditions would you expect to find dolomite stratigraphically above a coral reef facies? Below a coral reef facies?

Background Reading

Butler, G. P., 1969, Modern evaporite deposition and geochemistry of coexisting brines, the sabkha, Trucial Coast, Arabian Gulf: Jour. Sedimentary Petrology, v. 39, p. 70-89.

Pray, L. C., and Murray, R. C., eds., 1965, Dolomitization and Limestone Diagenesis, a Symposium: Soc. Economic Paleontologists and Mineralogists, Spec. Pub. 13.

Chapter Twelve **Stratigraphy** Patterns of Shoreline Sedimentation

Objectives

1. To examine in greater detail the depositional environments and associated sediments of shoreline and coastal areas.
2. To determine the stratigraphic patterns that will result from the lateral migrations or oscillations of these environments.
3. To determine the relationship of the resulting rock units to geologic time.

Important Terms

Transgression—A vertical stratigraphic succession in which the most offshore environments are highest, and the most onshore environments lowest.

Regression—A vertical stratigraphic sequence in which the most onshore sediments are highest, and the most offshore lowest.

Walther's Law—The idea that the vertical succession of rock types observed at a single locality reflects the lateral succession of environments that have migrated over that depositional site.

Time plane—An imaginary plane that cuts through a body of rock representing the actual topography on which sedimentation was taking place at one point in time.

Discussion

Reexamine the depositional environments shown in Figure 2.1. As an exercise, make a chart for each environment, listing the sediment characteristics (size, sorting, organic content), skeletal fossils likely to be preserved, trace fossils, and inorganic and organic sedimentary structures. Locate the same environments on Figure 12.1, and make a sketch map of this area showing their locations.

Consider now the effects of sea level changes on this environmental complex. If sea level were very gradually lowered, all of these environments would migrate in a seaward direction, but would maintain their general spatial relationships to each other. Eventually lagoon and barrier island sediments would cover over areas that had once been marine shelf (Figure 12.2). The lateral

South Toms River
Toms
River
Factory Branch
Deep Hollow
Swordens Pond
Cedar
Creek
North
Branch
Middle
Branch
Forked
River
Oyster
Creek
Lochiel
Creek
Sedge
Islands
BARNEGAT
BAY
ATLANTIC
OCEAN

Figure 12.1 Shoreline characteristics, Toms River, New Jersey

translation of these environments could generate relatively flat lying bodies of grossly uniform lithology bedded in vertical sequence. Such rock units are often named as *formations* if they are sufficiently thick and widespread. How many different formations could be named from the regressive sequence shown in Figure 12.2? What sequence of lithologies would be observed in a transgression, involving the same depositional environments?

Exercise

Part I

Rock samples or suites of rock samples collected from shoreline environments will be placed out in the laboratory. Identify the depositional environment of each, using all of the information at your disposal. Use the chart that you made above, comparing each sample suite to what would be expected in each depositional area.

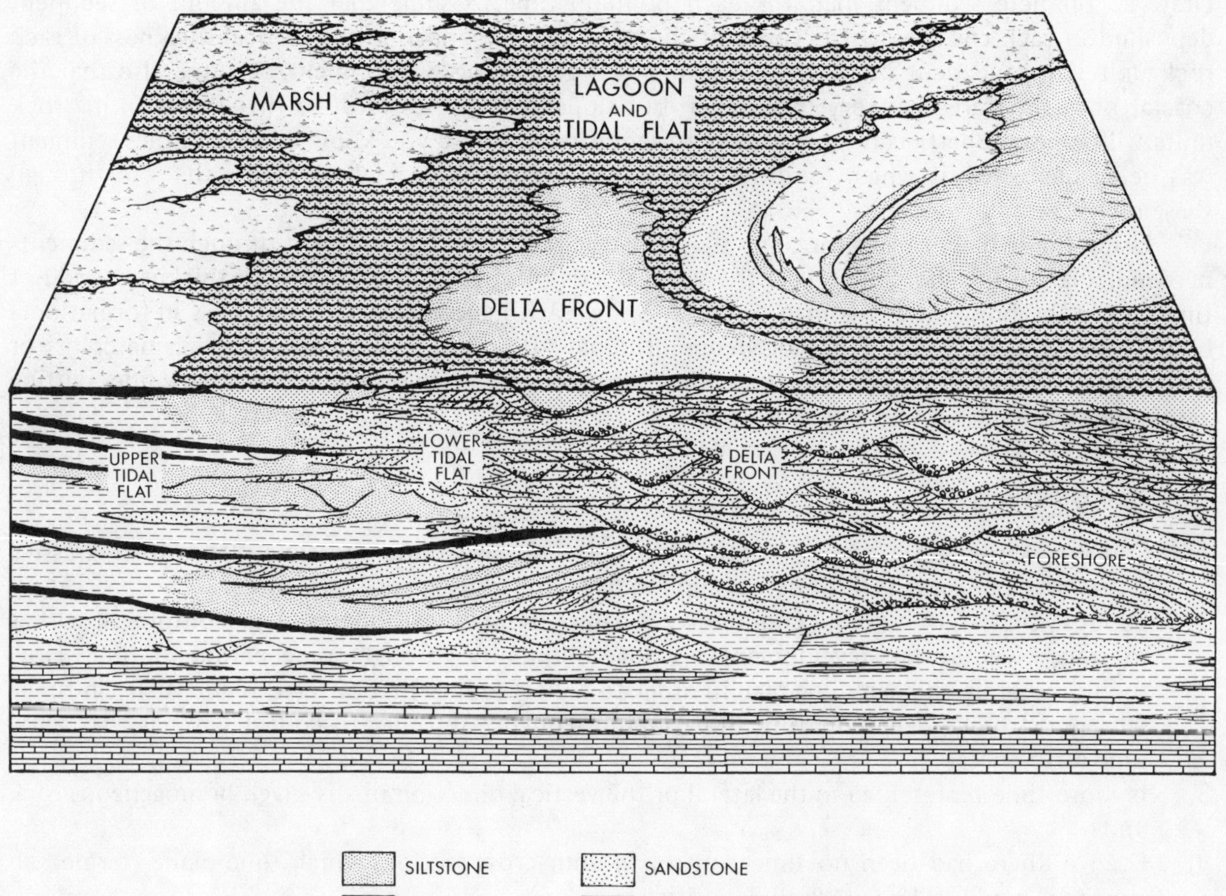

SILTSTONE
SHALE
COAL
LIMESTONE
SANDSTONE
SANDSTONE WITH GRAVEL
SANDSTONE, RIPPLED
SANDSTONE, CROSS-BEDDED

Figure 12.2 A regressive shoreline sequence

Part II

Using a full page of paper, draw a relatively straight shoreline 8 inches long at the left margin of the sheet. Using different colors or patterns, draw in the following environments from onshore to offshore, making each roughly parallel to the shoreline:

a. Lagoon-tidal flat, ¾ inch wide
b. Barrier island, ¼ inch wide
c. Marine shelf, detritals, 1 inch wide
d. Marine shelf, carbonates, extending to right edge of paper.

This exercise attempts to simulate environmental migrations that might take place over geologic time. Using sheets of tracing paper, this environmental pattern will be shifted in such a way as to generate both regressive and transgressive sequences. Assume that the pattern drawn on the base map represents time 0. On successive sheets of tracing paper, show the following changes:

a. Time 1—shoreline migrates 1 inch to the right
b. Time 2—shoreline migrates 2 inches to the right
c. Time 3—shoreline migrates 1½ inches to the left
d. Time 4—shoreline migrates 2 inches to the left

Draw a complete sediment map for each point in time. Assume that the amount of sediment deposited in each environment is constant throughout time, and that the vertical thickness of each rock unit is independent of the map width of its generating environment. Assume also that the coastal plain-shelf area has very little relief, but slopes slightly seaward, so that the resulting rock units will appear superficially horizontal in orientation. Because you know at which times sediment was being deposited at which localities, you can draw in the exact time planes that cut through these rocks.

Using the sediment maps for times 0-4, construct a vertical cross section through the sediments perpendicular to the shoreline. Make the vertical axis of your cross section rock thickness, and not time; make the horizontal axis map distance. Show each major rock unit, and draw in formational boundaries between units of homogeneous lithology. It will be helpful to use the same colors or patterns in the cross section as on the sediment maps. Draw in the time planes as dark lines cutting across the section; label time lines for times 0-4.

Test Questions

1. Can you formulate a general relationship between rock units and time planes for regressive shoreline sequences? Is this relationship modified for transgressive sequences?
2. Suppose that a given faunal community inhabits the shelf carbonates. A paleontologist wishes to trace evolutionary developments in this community through some period of time. How should he design his sampling plan in order to collect faunas that differed sequentially in time? What must be known about any stratigraphic sequence in order to collect samples that show time differences?
3. Is more time represented in the lateral or the vertical direction in any given homogeneous rock unit?
4. Even if there had been no time control in your cross section, which time plane (termed an *isochron*) could still be precisely identified?
5. Would the conclusions above apply to a barrier reef coastline equally well? With what modifications?

Background Reading

Dickinson, K. A., Berryhill, H. L., and Holmes, C. W., 1972, Criteria for recognizing ancient barrier coastlines, *in* Recognition of Ancient Sedimentary Environments, Soc. Economic Paleontologists and Mineralogists, Spec. Pub. 16, p. 192-214.

Eicher, D. L., 1968, Geologic Time, Prentice-Hall, Englewood Cliffs, N. J., p. 43-47.

Ferm, J. C., Milici, R. C., and Eason, J. E., 1972, Carboniferous Depositional Environments in the Cumberland Plateau of Southern Tennessee and Northern Alabama, Tenn. Div. Geology, Rept. Inv. 33.

Chapter Thirteen

Stratigraphy Patterns of Deltaic Sedimentation

Objectives

1. To examine in greater detail the depositional environments and associated sediments of marine deltas.
2. To determine the stratigraphic patterns that will result from the dynamics of deltaic sedimentation.
3. To determine the relationship of homogeneous rock units deposited by deltas to intervals of geologic time.

Important Terms

Delta—The locus of sedimentation where a stream issues into a larger body of water, either a lake or marine shelf.

Alluvial plain—Upper portion of a river's depositional area, where nonchannel deposition normally occurs only during floodstage.

Deltaic plain—Lower portion of a river's depositional area, near where it enters the sea; characterized by splitting of the main channel (*bifurcation*) into many separate *distributaries*.

Natural levees—Elevated ridges along the stream bank; caused by excess sedimentation along the stream margins during floodstage.

Meanders—Natural loops or sinuosity in the course of a stream channel.

Point bar—Accumulation of sand and silt on the inside of a meander loop.

Backswamps—Low lying, poorly drained areas, usually along the channel margins, separated from the channel by natural levees. A *crevasse-splay* deposit is formed when the natural levee is breached, and a miniature delta is built out into the backswamp area.

Distributary mouth bar—Deposit of sand and silt at the mouth of a distributary channel.

Discussion

This exercise considers the principal locus of detrital sedimentation, the place where streams dump their sediment load onto the marine shelf. Sediments are deposited in many environmental

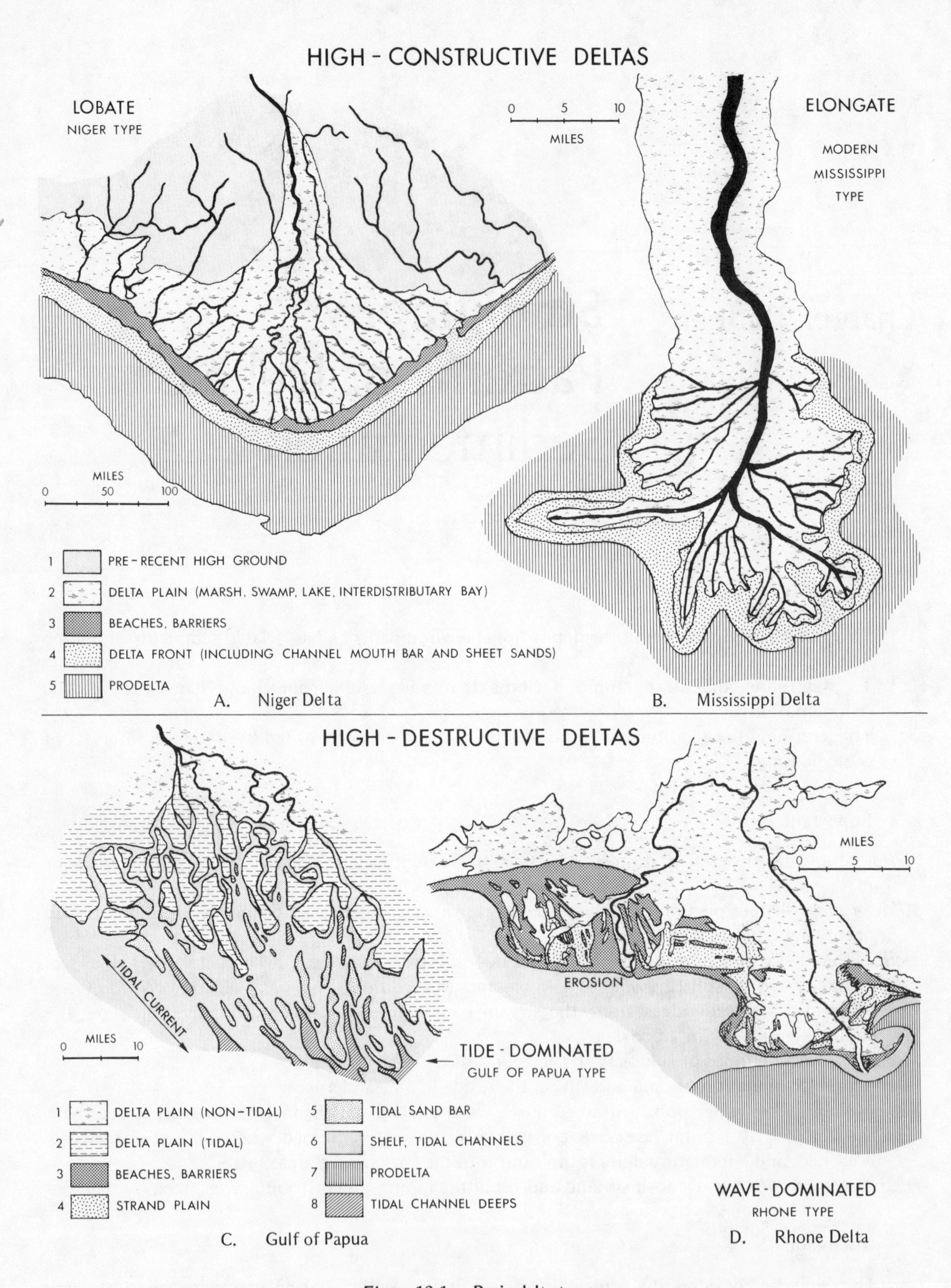

Figure 13.1 Basic delta types

complexes, including lakes, alluvial fans, ocean floors, and shorelines. The greatest volume of sediment, however, which produces the bulk of sedimentary rocks, is deposited in deltas and along shorelines. Deltas occur at the mouths of most major river systems; large amounts of sediment are dumped into the sea as the river water carrying the sediment loses velocity. The source of the sediment may be a large, generally flat drainage area, such as the interior region of the United States or Brazil, or a small, rugged, rapidly eroding mountainous region near the coastline. Nondeltaic shorelines (Figures 2.1, 12.1) occur where rivers bring much less sediment to the sea; this sediment is distributed down the coastline by currents. Deltas may be intensely modified by wave and current action (Figure 13.1) when the supply of sediment is low. In the Gulf of Mexico, the Appalachicola Delta (northern Florida) and the Rio Grande Delta (Texas) are highly modified by longshore currents, which distribute the sediment downcurrent along the coastline. The Mississippi River (Figures 2.2, 12.1), on the other hand, carries so much sediment that its delta is little modified by coastal processes. The Mississippi Delta provides a prototype for much of the discussion that follows.

Many of the sedimentational processes that characterize deltas are true of river systems in general, including processes producing meanders, point bars, natural levees, and swamps. Only the bifurcating distributaries, mouth bars, and adjacent marine waters are unique to the deltaic plain. On the alluvial plain, the river flows rapidly enough to maintain a single course, rarely overtopping its banks, and leaving the floodplain dry and unaffected by sedimentation except at times of abnormally high water. In the deltaic plain, in sharp contrast, the river is slowed by its entrance into the sea, and it can no longer carry its sediment load. The river then bifurcates into many separate channels in an attempt to cut through its own dumped sediment.

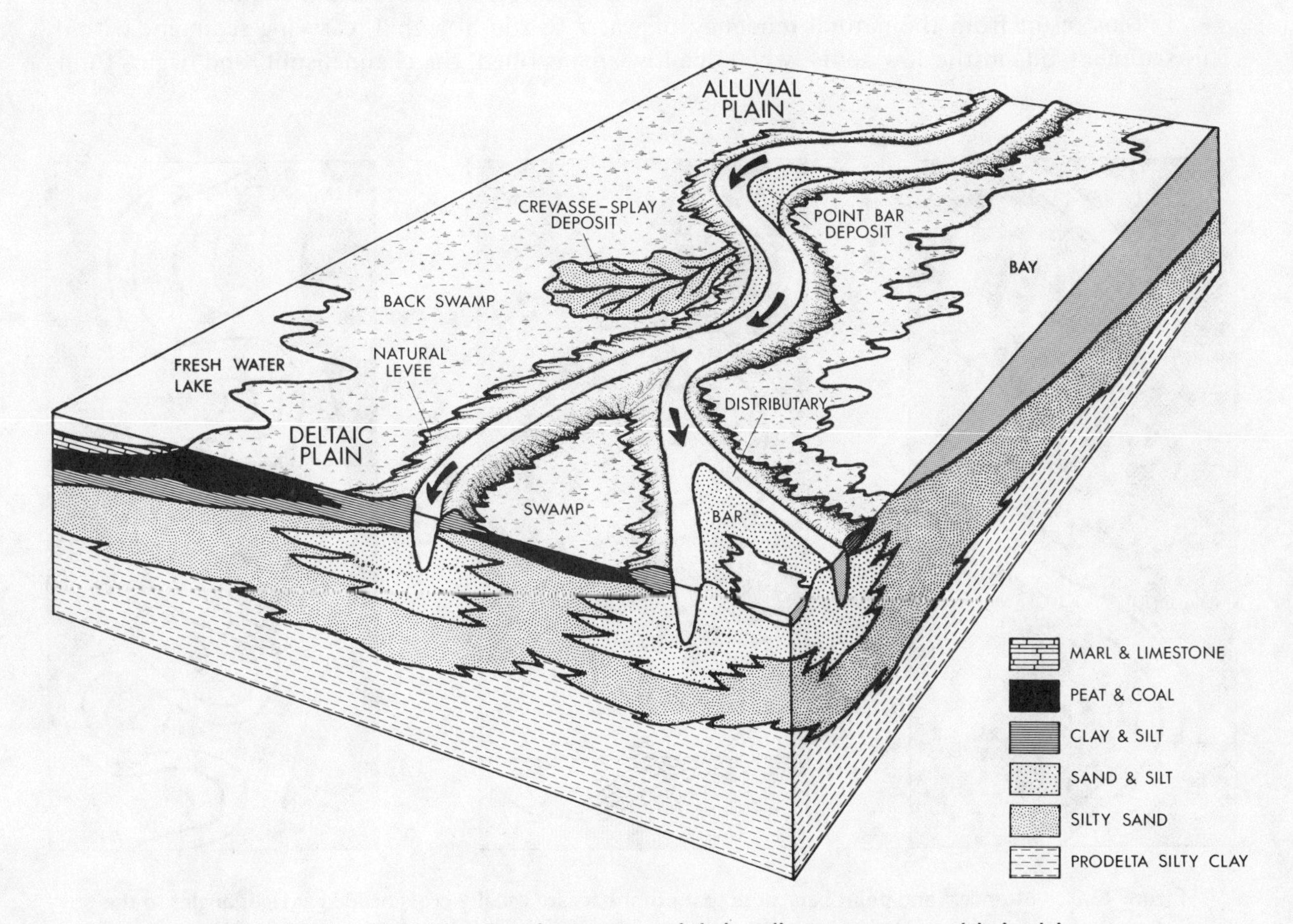

Figure 13.2 Depositional environments and their sediment types on a deltaic plain

Because sand is carried and deposited in faster moving water than silt and clay, sandy material is found nearer the channels, and muddy sediments farther away. Successive flooding of the channel causes the deposition of more sediment along the channel margins, forming natural levees (Figure 13.2), which eventually stand up above the surrounding area and act as barriers between the stream and the flood plain. The poorly drained backswamps will receive silt and clay near the channels, and may accumulate plant debris (to become coal) and carbonates (fresh water limestones) away from the channels.

The formation of point bar deposits by stream meanders is shown in Figure 13.3. Because the stream channel tends to maintain a constant cross sectional area, it cuts away material from the outside A′ bank as it fills on the inside A bank. As the cross section shows, the point bars are usually cross bedded at *right angles* to the direction of flow. Thus the entire channel constantly migrates, and may leave a series of natural levee ridges on the point bars. Eventually the stream may "cut off" the meander (Figure 13.5), and again take a more direct, straighter course. In doing so, it cuts through its own sediments again, producing an erosion surface. In stratigraphy, erosion surfaces in vertical sequences are called *unconformities*. The abandoned meander bend now has no sediment supply, except in times of flooding when it receives some silt or clay. Abandoned meanders filled with water become *oxbow lakes*.

The formation of distributary mouth bars is shown in Figure 13.4. Distributaries do not meander, but deposit sandy sediment principally at their mouths. Eventually the pile of sand at the mouth becomes very large, and the channel must split or bifurcate to get around the bar, forming two new distributaries which begin building their own mouth bars. In cross section (Figure 13.4), an area which has a muddy bottom receives coarser and coarser sediment as the channel advances, until the sandy bar itself arrives. These vertical sequences from fine to coarse sediment are called *fill-ins*.

Fill-ins result from the natural tendency of water to run downhill, carrying sediment with it. This sediment fills in the low spots; when one low area is filled, the channel splits and begins filling

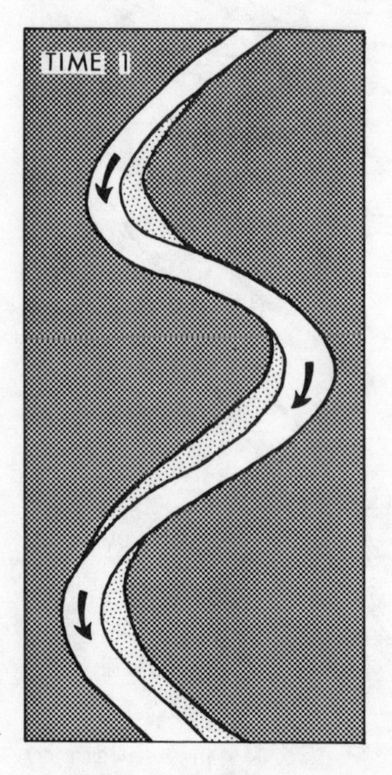

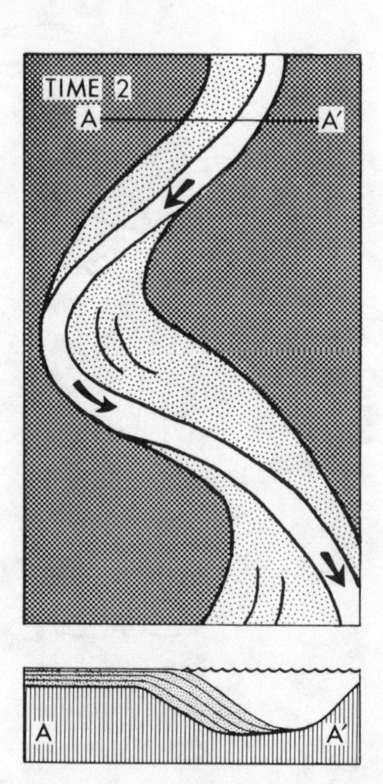

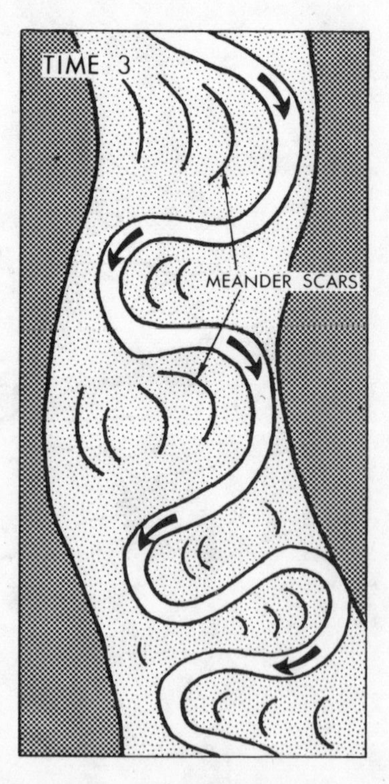

Figure 13.3 **Meanders and point bars; note that point bars are usually cross-bedded at right angles to the direction of flow.**

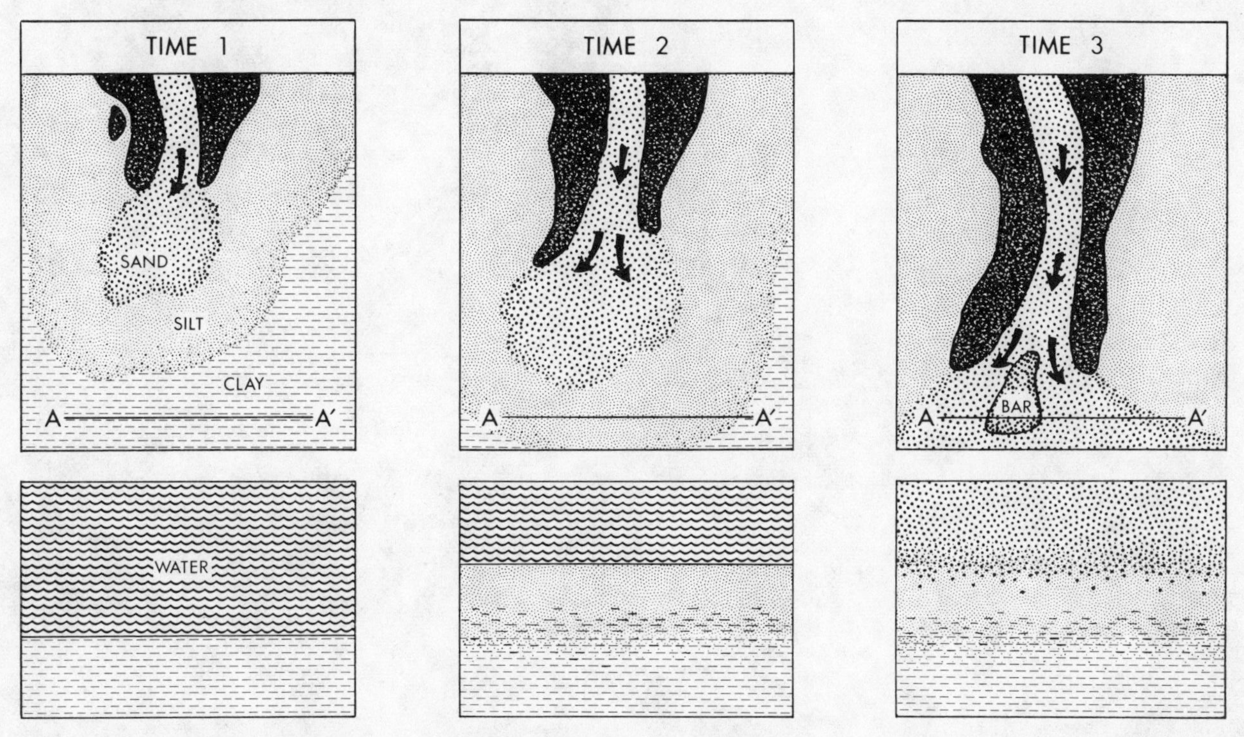

Figure 13.4 **Development of a distributary mouth bar, showing a vertical sequence from fine to coarse sediment.**

in another low area. This process takes place at many scales of magnitude in a delta, which is one of the most intriguing aspects of deltaic sedimentation. The smallest scale is that shown in the formation of distributary mouth bars; a single distributary may be only a few tens of feet wide. Many such channels (and backswamps) together combine to deposit a larger mass of sediment, such as the Cubits Gap sequence (Figure 13.6), which is a fill-in about 10 miles wide. The Cubits Gap complex is only one of eight major fill-ins that form the present Birdsfoot Delta, a mass of sediment about 40 miles wide, which began forming in the early 1800s (Figure 13.7A). The entire Mississippi Delta complex is a larger mass of sediment composed of at least seven such subunits, covering about 200 miles along the coast, and extending about 100 miles away from the alluvial plain. The development of this entire complex may represent several thousand years. Given several million years, fill-ins of the scale of magnitude of the Mississippi Delta complex could fill in very large areas of a continental shelf or inland sea (Chapter 17). The relationship of time planes to such a large body of sediment would indeed be complex.

Review now the sedimentary rock types forming in each of the environments discussed above (Figure 13.8). Two features of deltaic sedimentation which cannot be overemphasized are:

1. *The scale of variation*: Figure 13.8, as well as topographic maps displayed in the lab, show a deltaic area at some particular time. If you could walk across such an area, you would encounter a different sediment type every half mile to three miles. This means that no "bed," or unit of homogeneous lithology, is laterally continuous over a very large area at any one time, as is indicated in Figure 13.9. How can bodies of rock be formed that are larger than the environments in which they were deposited?

2. *The transitory nature of environments*: No depositional environment remains in one place very long, as seen in the examples of point bars and mouth bars. If you could stand in one place (as point A in Figure 13.8) for a hundred years or so, you could see sand, clay, plant debris, and many other sediment types accumulating at the same spot. A vertical column through the rocks formed at point A might be similar to that in Figure 13.9.

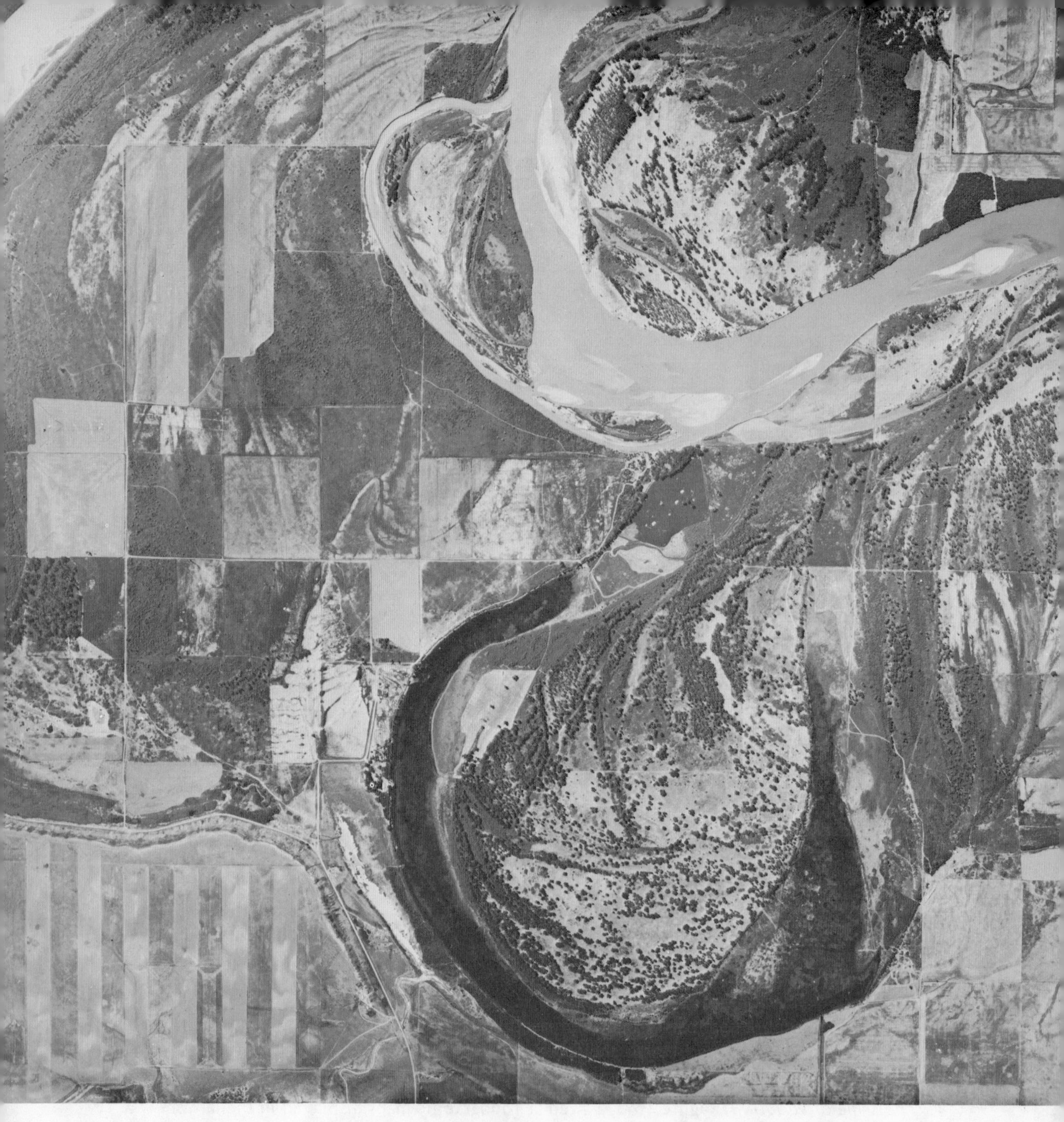

Figure 13.5 Aerial photograph of the Mississippi River's fluvial plain in Louisiana, showing a cutoff meander loop, oxbow lake, point bars, natural levees, and floodplain areas.

Exercise

Part I

Rock samples or suites of rock samples collected from deltaic environments will be placed out in the laboratory. Identify the sediment type, organic content, skeletal fossils, and sedimentary structures in each. Using all of the information at your disposal, identify as precisely as possible the most probable location of each unknown in Figure 13.8.

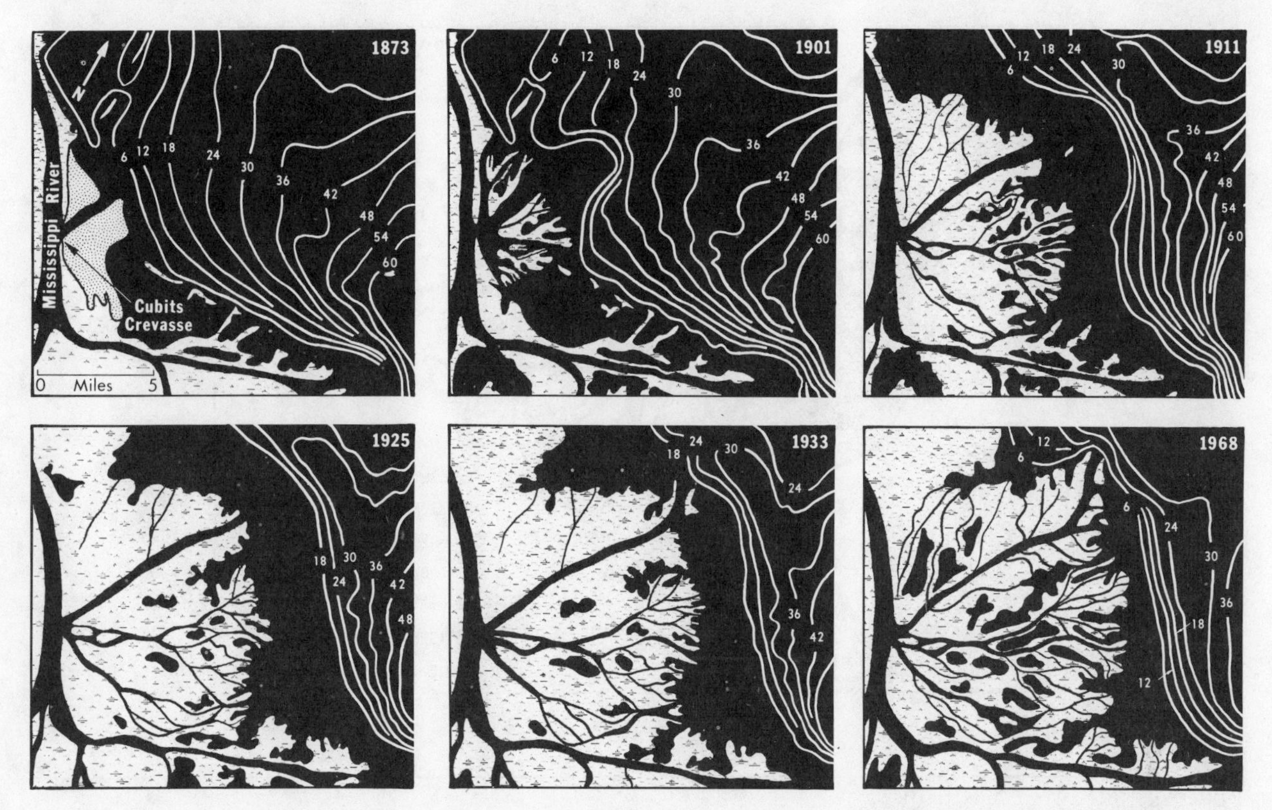

Figure 13.6 Sedimentary fill of Bay Rondo after the development of the Cubits Crevasse in 1860

Part II

Using some of the simple descriptions of the behavior of deltaic depositional environments, you can now "build" a three-dimensional volume of deltaic sediment, roughly similar to the Cubits Crevasse fill-in (Figure 13.6). The worksheet (Figure 13.11) shows a map of a hypothetical deltaic plain area at a beginning time, time 0. The map shows channels, swamps, bars, and their associated sediment types, using the same symbols as in Figure 13.8 and 13.9. The sequence of sedimentary events is given below for successive points in time. Using the sheets of tracing paper, redraw the sediment map for each time interval showing the spatial position of all sediment types.

All areas on the map will be receiving some type of sediment at any one time. By superimposing the sediments deposited at different times, you can generate a cross section through the sediment mass. You may choose the location of the AA′ traverse; fill in the cross section for each time period as you draw the maps. Remember that sediments are piled up horizontally on top of each other, and that time planes are not horizontal, but follow the topography of the surfaces on which sediment is accumulating.

Time I: Distributary A builds seaward ½ mile and bifurcates, forming channels F and G.
Distributary B builds seaward and combines with C to form E.
Distributary D builds seaward ½ mile.

Time II: Distributary F builds its new channel ½ mile seaward; G builds seaward ½ mile and bifurcates, forming channels H and I.
Distributary E builds seaward ¼ mile and bifurcates into channels J and K.
Bay Y (which was enclosed when B and C joined) is now a swamp; fill it with plant debris, clay, or marl.
Distributary D builds seaward ½ mile and bifurcates into channels L and M.

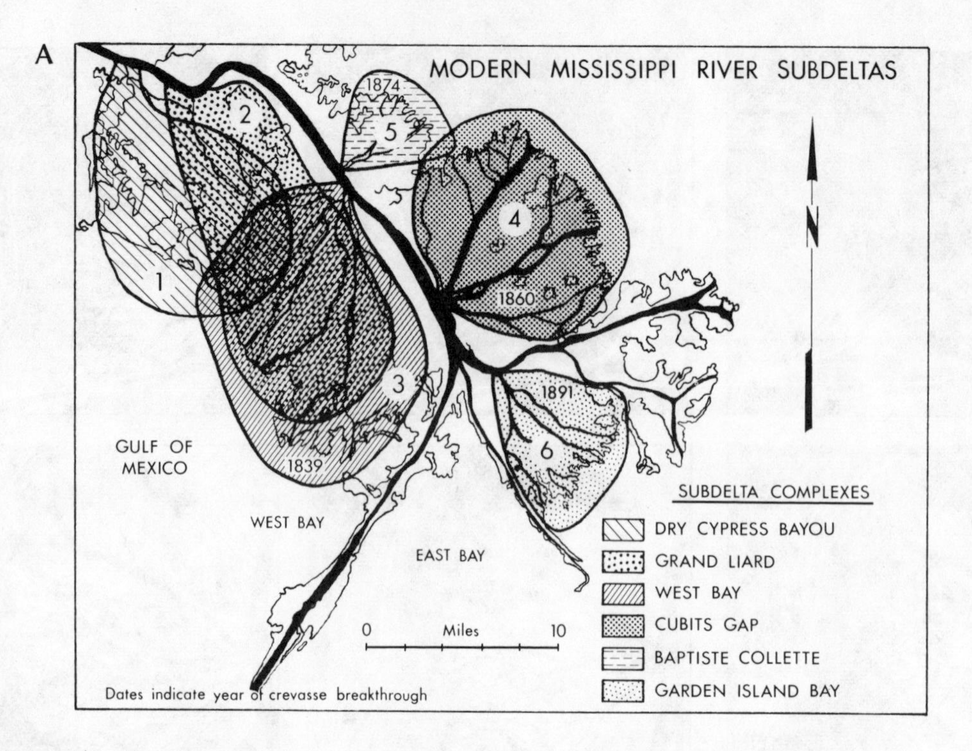

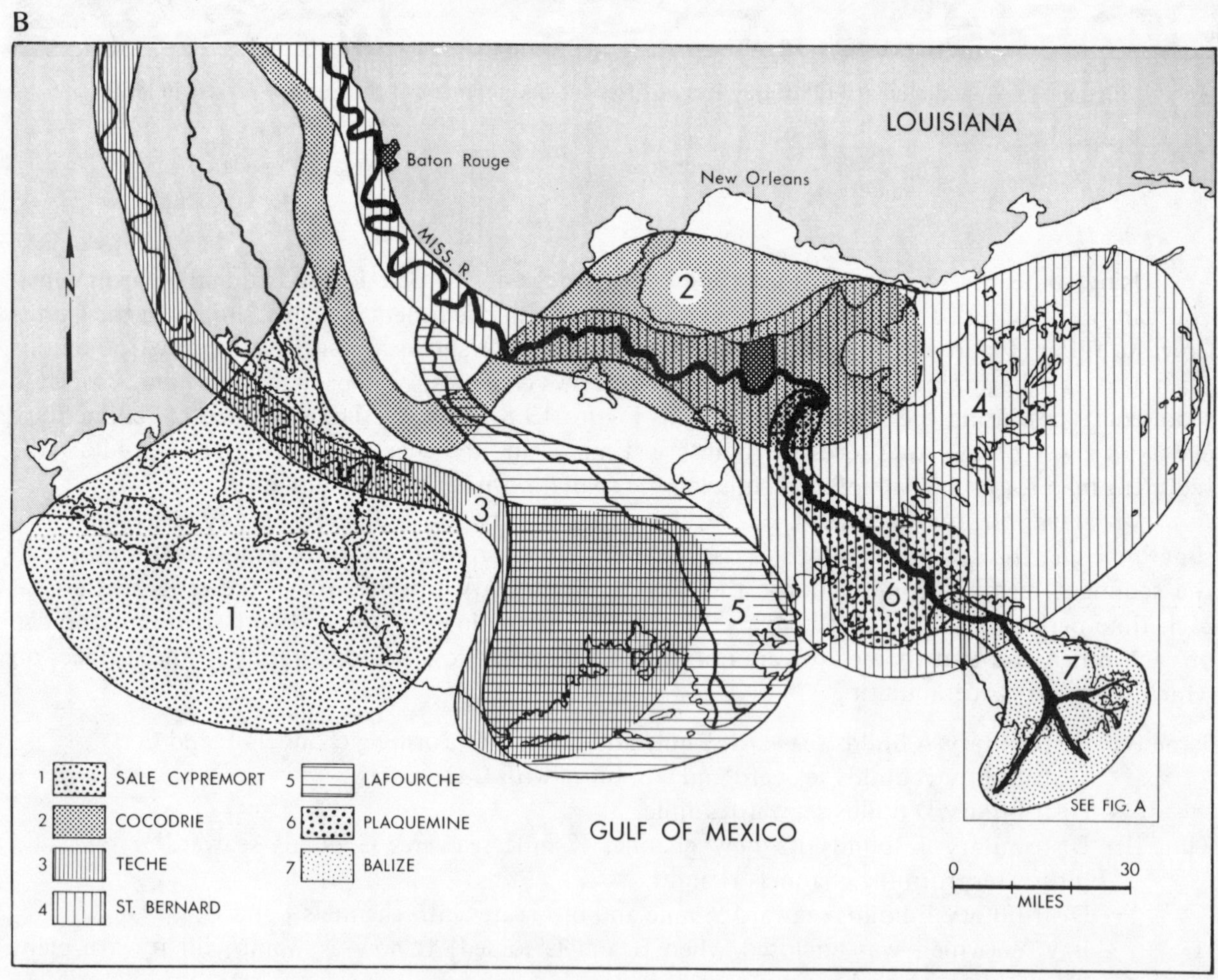

Figure 13.7 Subdeltas of the present Birdsfoot Delta (A), and older deltaic complexes (B).

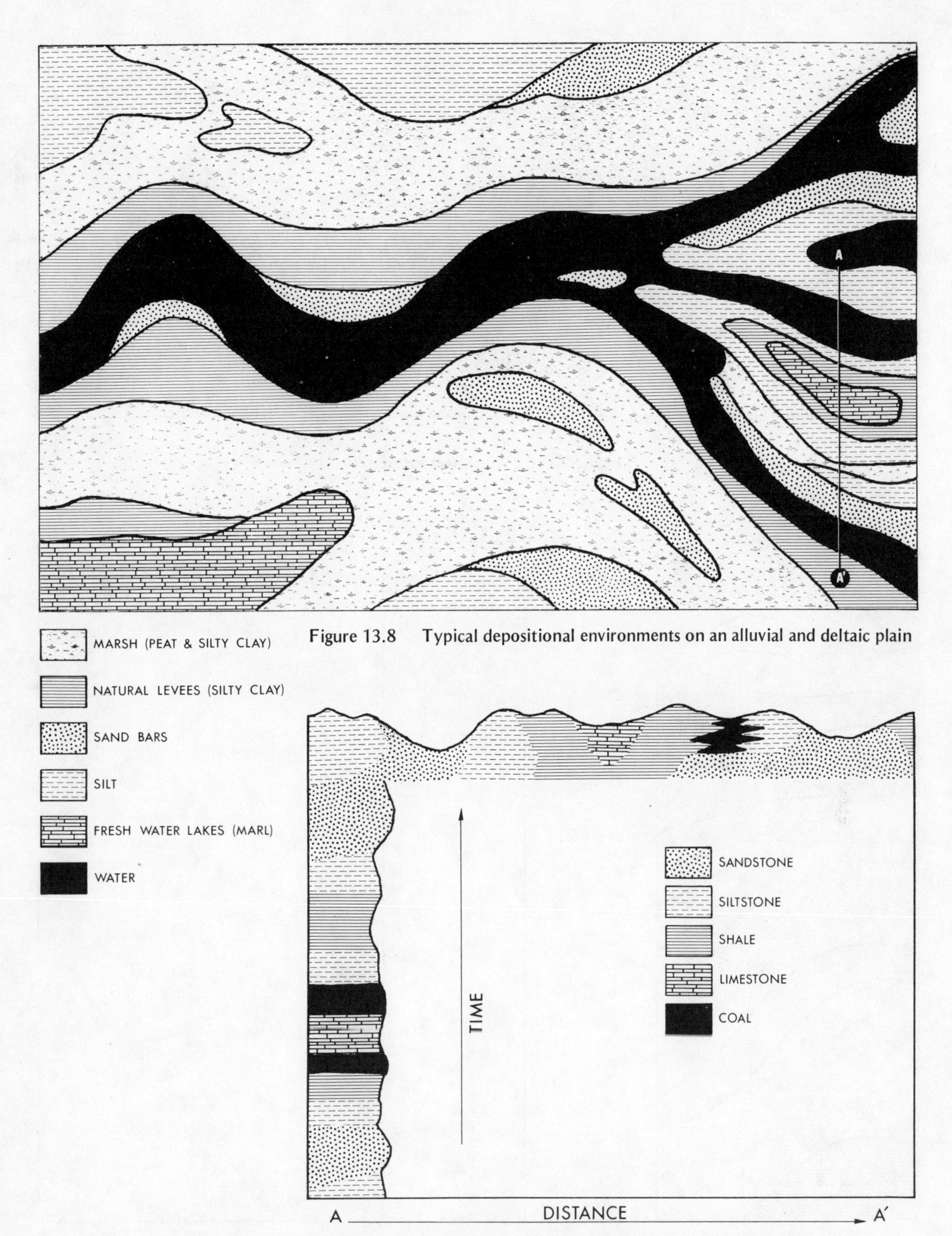

Figure 13.8 Typical depositional environments on an alluvial and deltaic plain

Figure 13.9 Sedimentary rock types that would result from the lateral sequence of environments shown along traverse A-A′ in Figure 13.8.

Figure 13.10 Worksheet for exercise

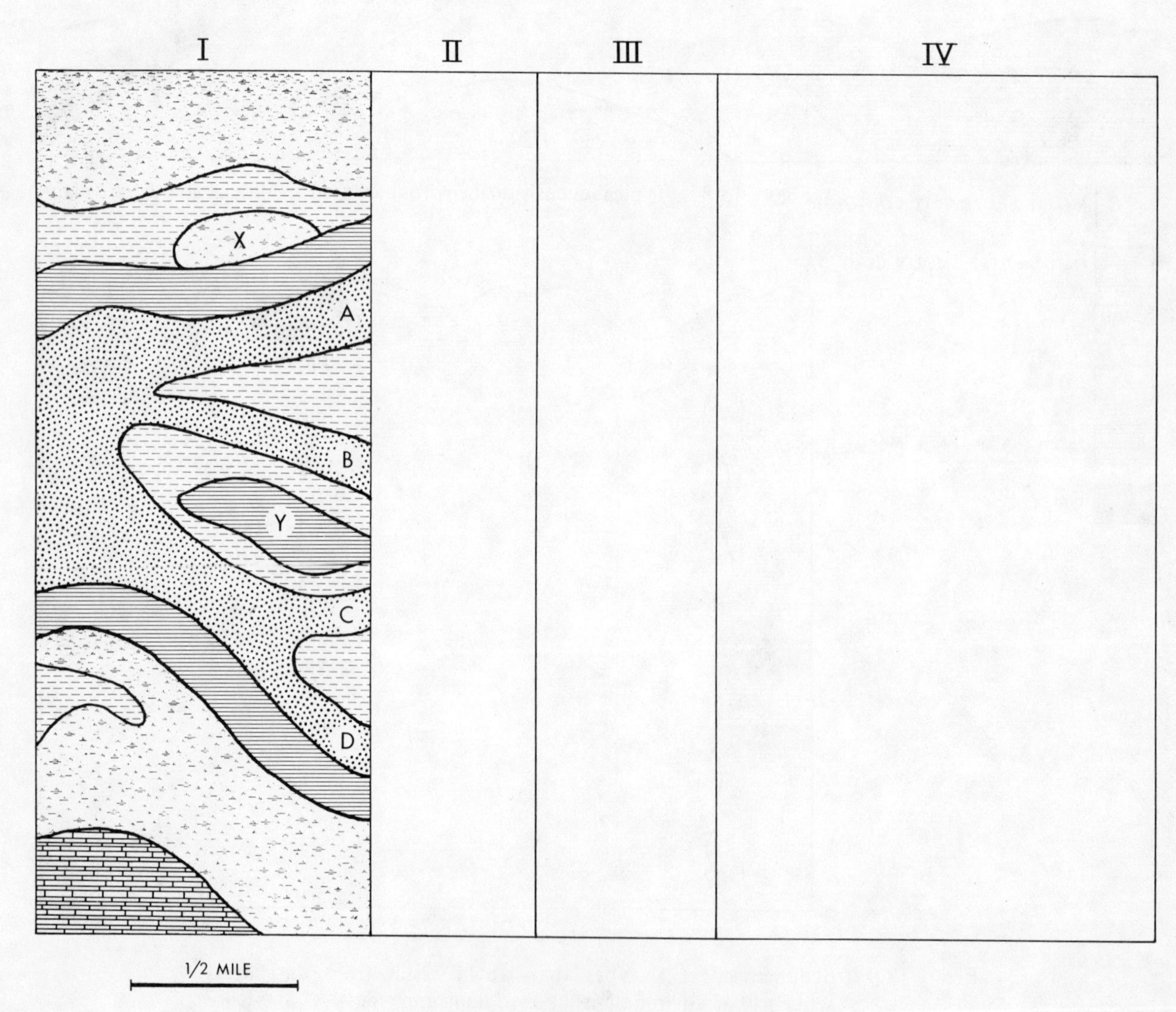

Figure 13.11 Worksheet for exercise

Time III: Distributaries F, H, and I build seaward ½ mile.
Distributary J builds seaward ¾ mile and bifurcates.
Distributary K builds seaward ½ mile and bifurcates.
Distributary L builds seaward ½ mile.
Distributary M is closed at the branch point.

Test Questions

1. Carboniferous geologists have observed that the same sequences of coal, shale, sandstone, and limestone may be observed over and over again in the inland sea deposits of the United States. If these are deltaic sequences, what causes this apparent cyclicity? Does it mean that the sea completely advanced and retreated over the entire region one time for each observed cycle?
2. Will an erosion surface always be present below a point bar deposit? Could the process of point bar sedimentation produce a laterally continuous erosion surface in the resulting rocks? Explain your answer.
3. What are the differences in three dimensions between point bar sandstones and mouth bar sandstones? What would be the geometry of a mouth bar sandstone if the channel kept extending itself for several miles before bifurcating? What feature might be observable on the upper surface of such a sandstone body?
4. Does the direction of cross bedding measured over a large region reflect the directions of water flow and sediment transport? Explain your answer.
5. What is the relationship between time planes and homogeneous rock units in a body of sediment such as the Cubits Crevasse complex? What is the relationship between time planes and the larger subunits of a deltaic complex?
6. In deltaic sediments, does the cyclic repetition of sediment types in numerous cycles at many widely separated localities mean that each cycle (in sequence) represents a simultaneous, regionally widespread, sedimentary event? Explain your answer.

Background Reading

Morgan, J. P., ed., 1970, Deltaic Sedimentation, Modern and Ancient, Soc. Economic Paleontologists and Mineralogists, Spec. Pub. 15.

Ferm, J. C., Horne, J. C., Swinchatt, J. P., and Whaley, P. W., 1971, Carboniferous Depositional Environments in Northeastern Kentucky, Geol. Soc. Kentucky, Spring Field Conference.

Smith, G. E., Dever, G. R., Horne, J. C., Ferm, J. C., and Whaley, P. W., 1971, Depositional Environments of Eastern Kentucky Coals, Kentucky Geol. Survey, Univ. of Kentucky.

Chapter Fourteen

Glacial Geology
Patterns of Glacial Sedimentation

Objectives

1. To examine in detail the sediments and depositional environments associated with the margins of continental ice sheets.
2. To determine the stratigraphic patterns that result from the dynamics of continental glacial sedimentation.
3. To determine the relationship of homogeneous sedimentary units deposited by glacial processes to intervals of geologic time.

Important Terms

Till—Poorly sorted clay, silt, sand, and pebbles formed by glacial erosion and reworking of older materials; twigs and branches are common in the lower portions of a till when glacial lobes have overridden ancient forests.

Outwash—Well-sorted sand and gravel deposited as point bars on braided streams flowing out of a melting glacial lobe; deposit is a fan-shaped wedge thinning away from the tip of the lobe.

Loess—Wind-deposited clay and silt blown off of outwash plains, due to glacial disruption of vegetational cover; loess and sand dunes are prominent on the side of the outwash plain away from the prevailing wind direction; loess thins away from the source of the material.

Lake (lacustrine) deposits—Silt and mud deposited in lake water dammed up behind a terminal moraine (see below) or by an ice mass; in terminal stages of lake development, peat and plant debris can accumulate. Lake silts are often *varved* due to annual turnover of the water. Sand dunes often form on beach areas away from the prevailing wind direction; such sandy beach-dune deposits resemble marine shoreline sediments. Glacial lake sediments often contain numerous snails and other shells, forming *marl*.

Terminal moraine—Ridge of glacial till pushed up by the oscillations of an ice lobe pushing against its own sediments.

Glacial lobe—Tongue-shaped ice mass associated with the margins of a continental glacier, caused by pressure differences in the ice sheet as well as the topography of the proglacial surface.

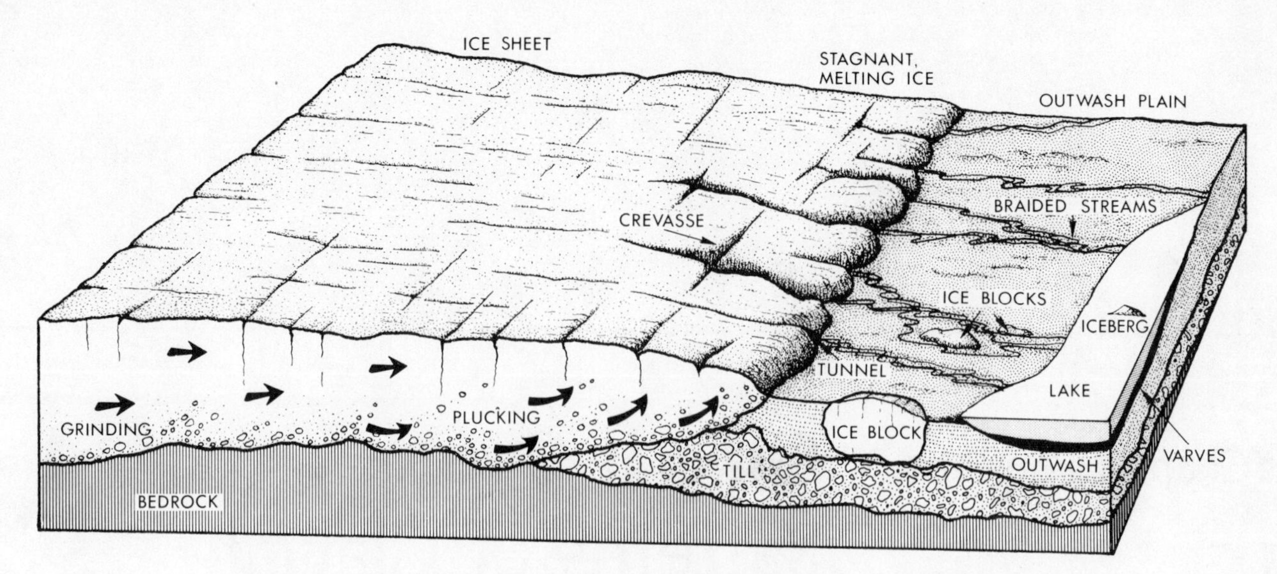

Figure 14.1 Physical processes and sediments associated with a stagnant ice front

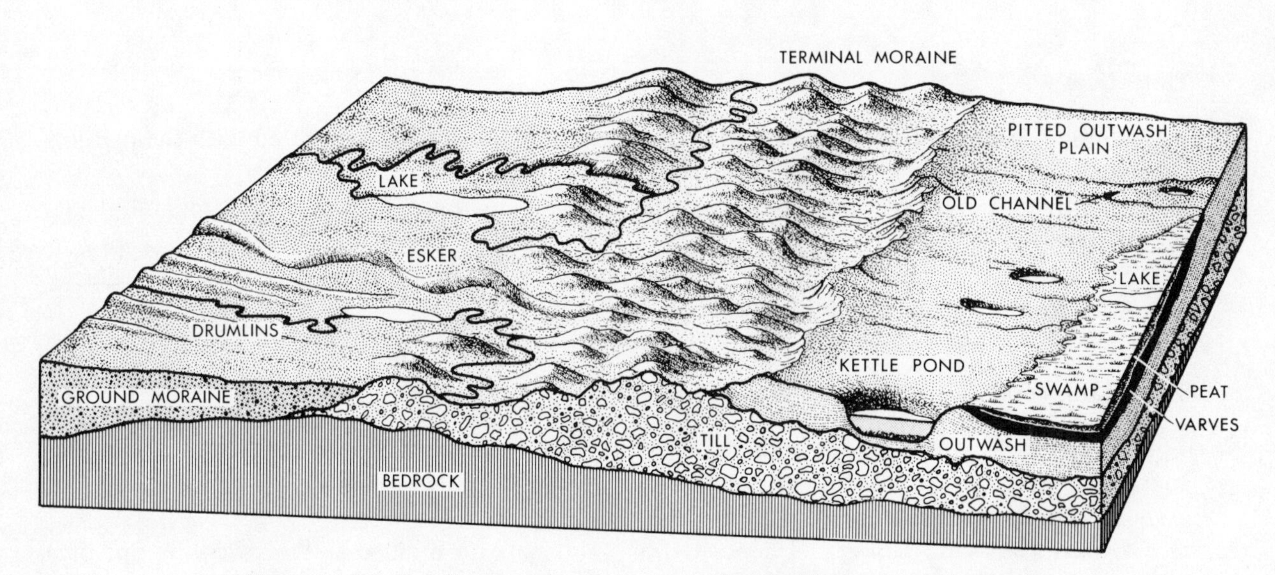

Figure 14.2 Landforms and sediments associated with a retreated ice front

Figure 14.3 Glacial map of parts of Michigan, Ohio, and Indiana.
Blue—Lake sediments
Yellow—Outwash
Orange—Ice-contact stratified sediments
Dark Green—Terminal moraines
Light Green—Till plains
Pink—Older till plains
Gray—Unglaciated areas
Blue lines—Striation grooves
Red lines—Beach deposits

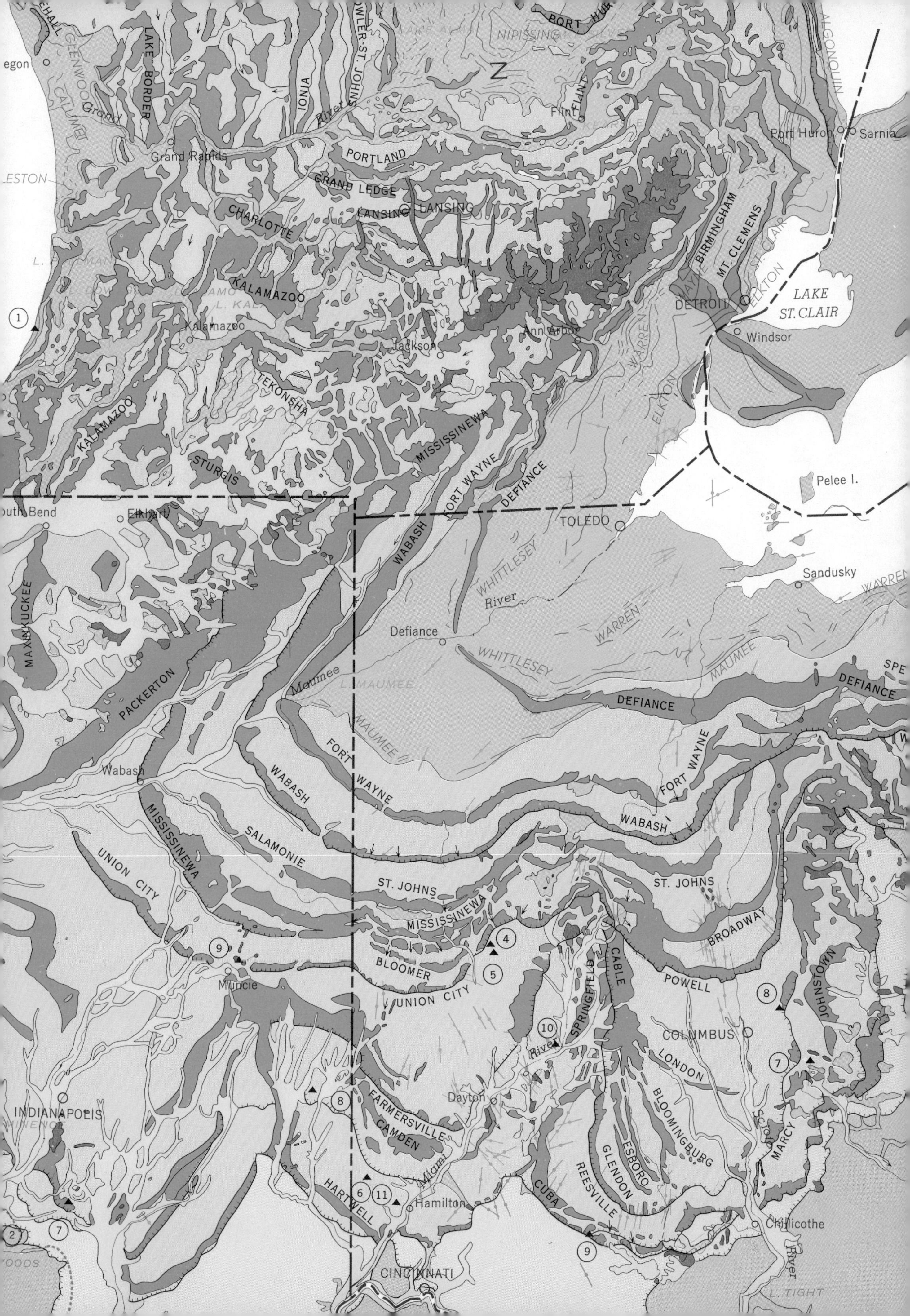

LAKE BORDER
IONIA
PORT HURON
LAKE ALMA
NIPISSING
FLINT
Flint
Port Huron
Sarnia
ALGONQUIN
Grand
River
Grand Rapids
PORTLAND
GRAND LEDGE
LANSING
Lansing
CHARLOTTE
KALAMAZOO
BIRMINGHAM
MT. CLEMENS
DETROIT
ELKTON
LAKE ST. CLAIR
Windsor
Kalamazoo
Ann Arbor
Jackson
WARREN
TEKONSHA
MISSISSINEWA
FORT WAYNE
DEFIANCE
STURGIS
Pelee I.
Elkhart
TOLEDO
WABASH
MAXINKUCKEE
WHITTLESEY
River
Sandusky
Defiance
PACKERTON
Maumee
L. MAUMEE
MAUMEE
Wabash
SALAMONIE
UNION CITY
ST. JOHNS
BROADWAY
BLOOMER
POWELL
Muncie
CABLE
SPRINGFIELD
COLUMBUS
JOHNSTOWN
LONDON
INDIANAPOLIS
Dayton
FARMERSVILLE
CAMDEN
BLOOMINGBURG
ESBORO
GLENDON
REESVILLE
CUBA
MARCY
HARTWELL
Miami
Hamilton
Chillicothe
CINCINNATI
L. TIGHT

Figure 14.4 Topography developed by continental glaciation, showing terminal moraine, outwash plain, and till plain.

Discussion

Continental ice sheets covered much of the northern hemisphere during the Late Cenozoic. Earlier glaciations have been proposed to explain sedimentary sequences in the Permian of the southern hemisphere, the Ordovician of the Sahara region, and the Late Precambrian of the Great Lakes area of North America. Of these, the evidence for the Cenozoic glaciation is the most complete and undisputed.

The effects of mountain (alpine) glaciation are not included here, because they do not compare with continental ice sheets either in amount of sediment generated or in the length of time their sediments will remain uneroded. Continental glaciers, like deltas, commonly deposit sediments in local, lobate masses. Continental glacial margins form large, lobate tongues that advance and retreat locally in time. Such features may have been 100 to 200 miles long, and 50 to 100 miles wide. The present-day shape of Lake Michigan reflects the geometry of a Pleistocene glacial lobe; the melting of this ice tongue may have provided much of the water needed to fill the lake. The continental glacial deposits of the interior United States display a complex of highly lobate patterns (Figure 14.3). Like the distributary complexes of a deltaic plain, the sediments deposited by a glacial lobe also represent a particular configuration of depositional environments.

Glacial action consists of eroding and redistributing older sediments and bedrock by scouring off the areas over which the ice moves, and transporting the sediment by ice flow to the margins of the ice sheet, where melting and sedimentation occur (Figures 14.1, 14.2).

Glacial lobes advance due to pressure in the ice mass caused by the weight of the ice "upstream." When the pressure is alleviated by ice flow, the advance ceases. If ice no longer feeds an advancing lobe, it begins to melt and retreats by melting. As the ice melts and retreats, the sediment incorporated into it is thereby released and deposited as till, outwash, loess, or lake deposits. A terminal moraine (Figure 14.4) is formed by a lobe pushing up a ridge of sediment as it advances. Both glacial lobes and outwash plains can bury preglacial grasslands and forests. The winds created by meteorological conditions along a glacial margin will rework the loose outwash (not yet stabilized by plant cover) into sand dunes and loess.

Ice, like a very viscous liquid, generally flows downhill. Consequently, once a low-lying area has been filled in by glacial sediments, newly formed lobes advancing from the ice front will seek out other areas. Alternatively, a lobe could advance over the same area several times (Figure 14.5) until it has been aggraded enough to cause new lobes to flow elsewhere. Thus glacial sedimentation is somewhat analogous to the *fill-ins* of deltaic sedimentation.

Exercise

Part I

Sediment samples or sample suites collected from different glacial environments will be placed out in the laboratory. Identify the sediment type, organic content, and any fossils (plant or animal) in each known and unknown sample. Identify the most likely source of each sample in Figure 14.6.

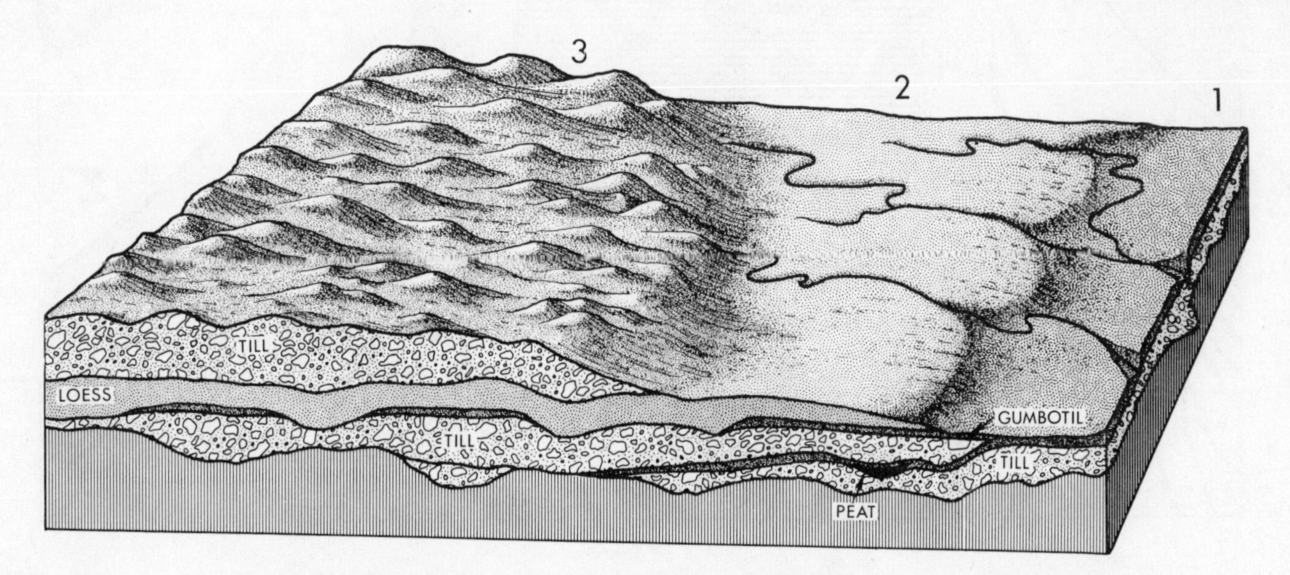

Figure 14.5 Succession of sediments associated with three advances of a glacial lobe

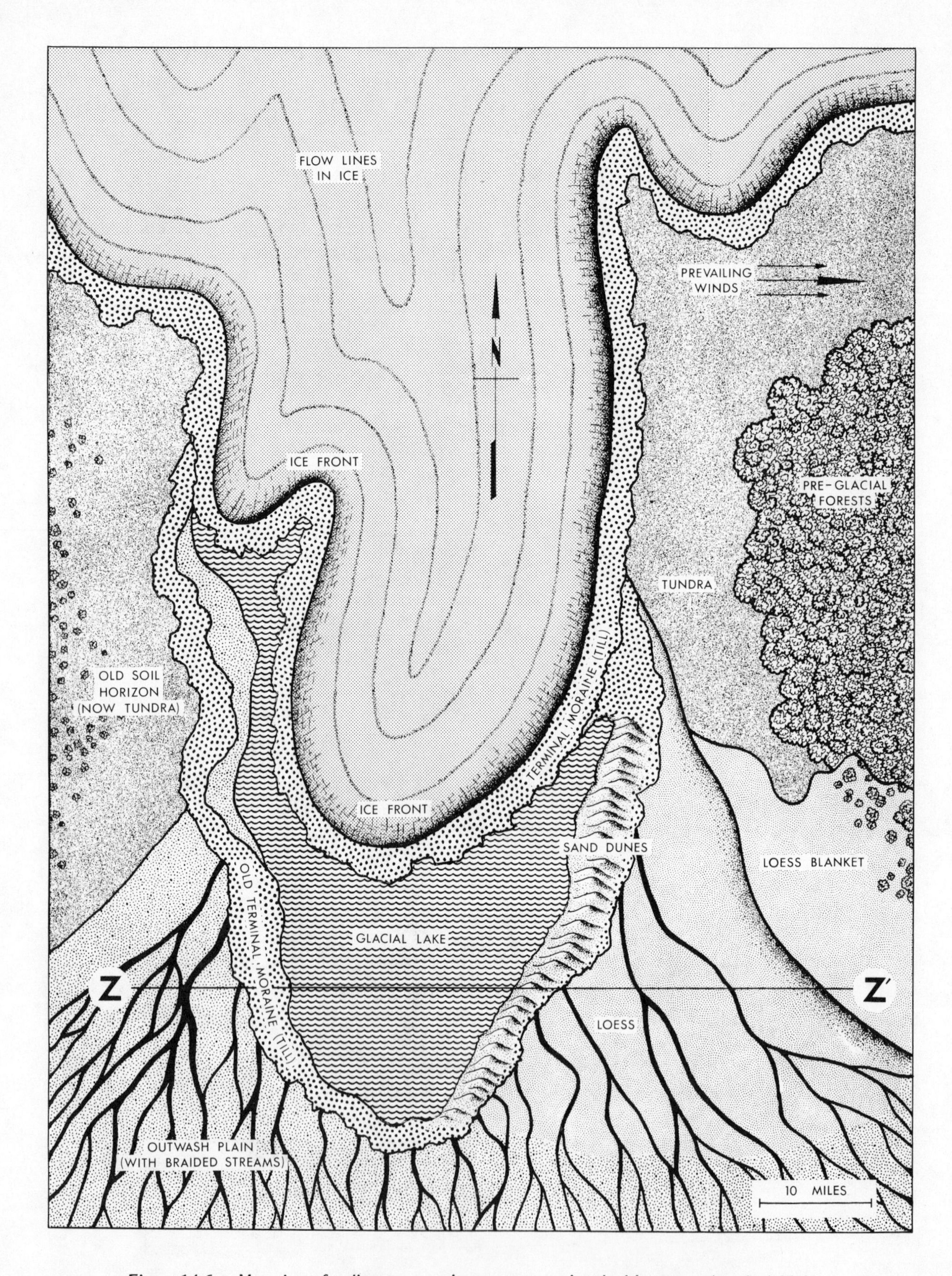

Figure 14.6 Map view of sedimentary environments associated with a retreating glacial lobe

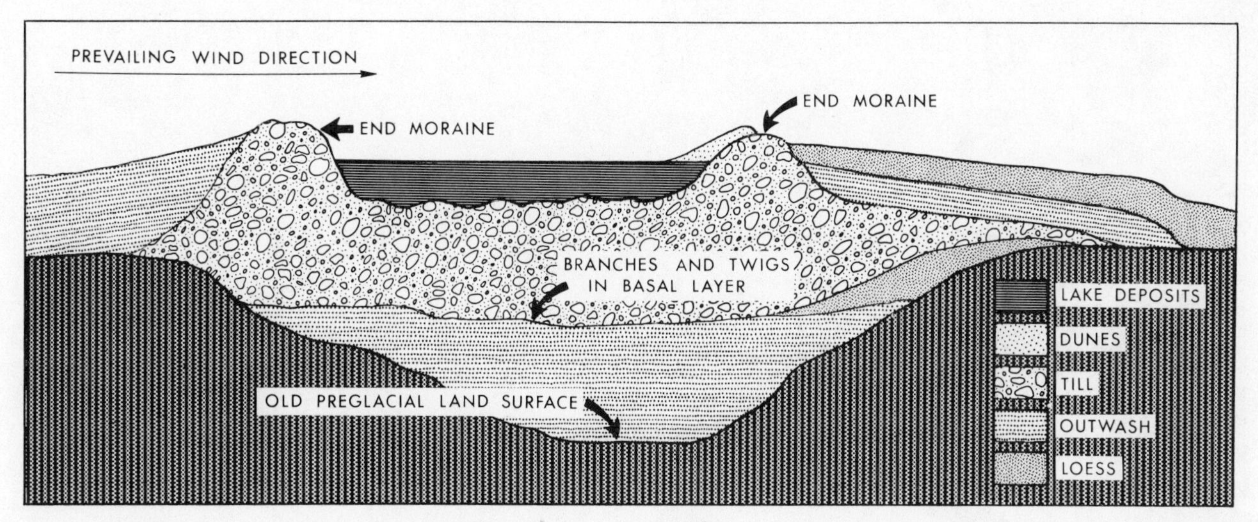

Figure 14.7 **Cross section (Z-Z') of sediments for Figure 14.6, with extreme vertical exaggeration**

If time permits, examine the sorting of one or more glacial tills by washing the sediment with water through a stack of nested sieves. Concentrate the sediment in a corner of the sieve, and wash it into a beaker, using only a minimal amount of water. The beakers may be dried either in the air or in a drying oven. Weigh each fraction, and present the results as a histogram (bar graph). Compare these to the sediments studied in Chapter 1.

Examine only the sand fraction of the glacial till, and determine the roundness (Figure 1.1) of the grains. Compare these with the sand dune material and the outwash. What different effects do wind, ice, and water have on grain roundness?

Part II

Figure 14.6 shows a single lobe of a hypothetical continental glacier. Figure 14.7 illustrates a cross section of the sediments that would be found along the traverse Z-Z′. Figure 14.8 shows the ice front of a continental glacier at time 0. Lobes will advance and retreat from this glacier at particular times and deposit all of the sediments shown in Figures 14.6 and 14.7. Your assignment is to generate this glaciation according to the following steps, and draw cross sections through the pile of accumulated sediment along the traverses X-X′ and Y-Y′. All time planes and sediment types must be shown in the cross sections. Draw sediment maps for time 0, and each of the following, using the overlay sheets:

Time I: Lobe C advances 300 miles south.
Lobe D advances 200 miles southeast.
Lobe A retreats 100 miles northeast.

Time II: Lobe C advances 100 miles south.
Lobe D advances 100 miles southeast.
Lobe A advances 300 miles southwest.
Lobe B advances 100 miles south.

Time III: Lobe C retreats 200 miles north.
Lobe A advances 200 miles southwest.
Lobe B advances 100 miles south.

Time IV: Lobe C retreats 300 miles north.
Lobe D retreats 200 miles northwest.
Lobe B advances 100 miles west.

Time V: All ice retreats from area.

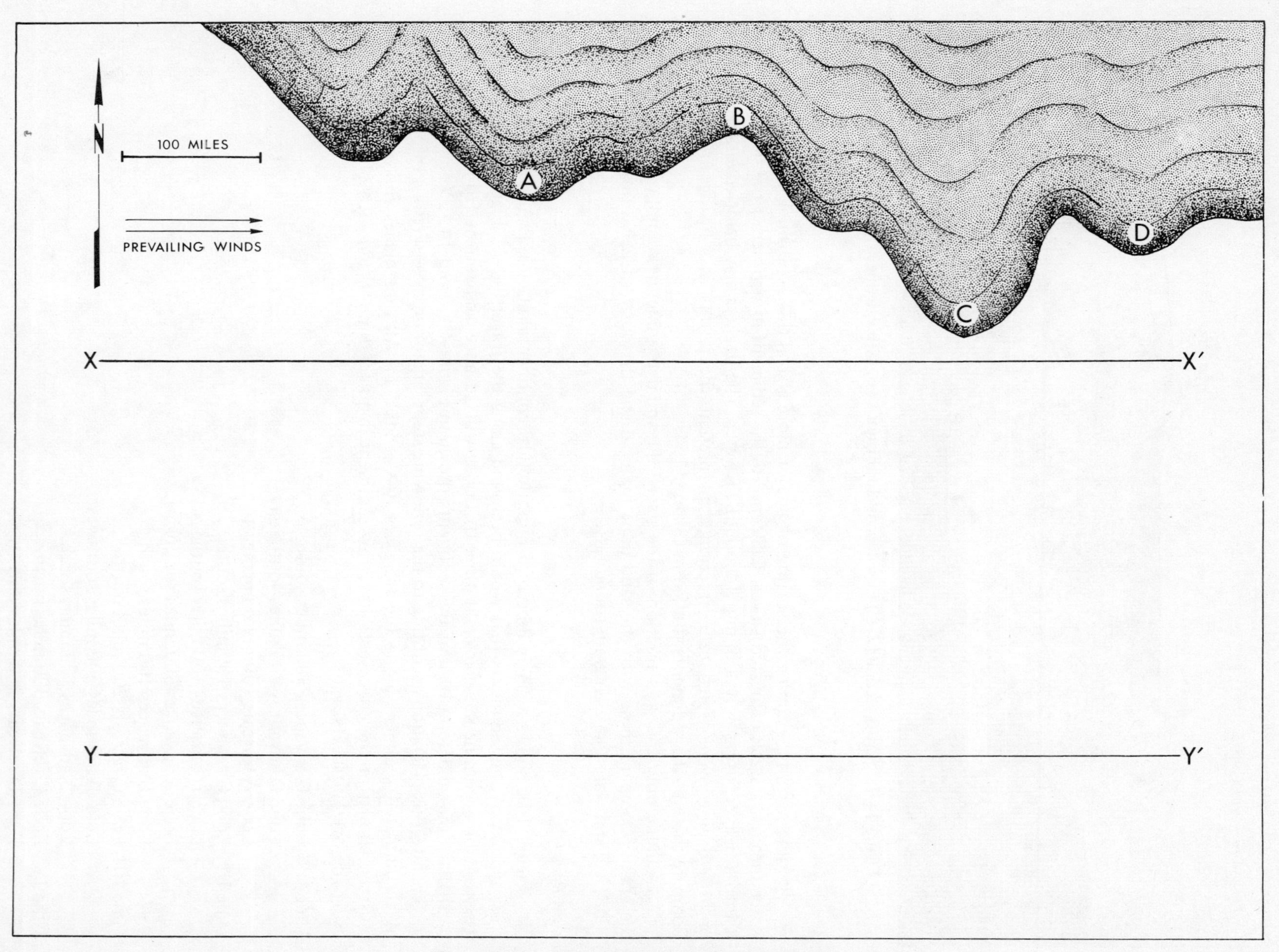

Figure 14.8 Worksheet for exercise—map view

X X′

Y Y′

Figure 14.9 Worksheet for exercise—cross sections

Test Questions

1. In glacial sediments, does the cyclic repetition of sediment types in numerous cycles at many widely separated localities mean that each cycle (in sequence) represents a simultaneous, regionally widespread sedimentary event? Explain your answer.
2. What is the general relationship between time planes and homogeneous sediment units in glacial materials?
3. In what areas around an ice sheet could a plant cover and soil horizons develop? What is the time relationship between such areas and those being actively glaciated?
4. Certain ancient sedimentary rocks have been interpreted as glacial tills and outwash, based largely upon poor sorting, striated (scratched) pebbles, and striated bedrock surfaces underlying the sediments. Could these sediments have been deposited in any other sedimentary complex, such as an alluvial fan? Would there be any difference in the resulting stratigraphic patterns?

Background Reading

Flint, R. F., 1973, The Earth and its History. W. W. Norton and Co., New York, p. 307-51.

Hamilton, W., and Krinsley, D., 1967, Upper Paleozoic glacial deposits of South Africa and southern Australia: Bull. Geol. Soc. America, v. 78, p. 783-800.

Strahler, A. N., 1972, Planet Earth: its Physical Systems through Geologic Time, Harper and Row, Publishers, New York, p. 204-209.

Chapter Fifteen

Geotectonics
Tectonic Sedimentary Patterns

Objectives

1. To examine in greater detail sediments associated with the local deformation of the earth's crust.
2. To determine the stratigraphic pattern that will result from the dynamics of tectonically controlled sedimentation.
3. To determine the relationships of homogeneous sedimentary units deposited by tectonic processes to intervals of geologic time.

Important Terms

Block fault—A high angle fault, causing the block on one side to be upthrown and that on the other side to be dropped down with respect to each other.

Graben—An elongate sedimentary basin formed by a downdropped block bounded on both sides by high angle faults.

Bajada—A continuous blanket of sediment formed by coalesced alluvial fans along a fault escarpment.

Discussion

The preceding three chapters have illustrated the relationships between geologic time and homogeneous rock units for shoreline, deltaic, and glacial sedimentation. This exercise will complete the picture by considering the last major environmental complex, the alluvial fan-bajada type of basin filling. In the western United States, almost all of the intermontane basins experienced this type of sedimentation during the Tertiary Period. In addition, the Triassic block fault valleys of the eastern United States (Figure 15.1), as well as Precambrian basins in Montana (the LaHood facies of the Belt) represent sedimentation of this type. Alluvial fan basin filling characterizes the early stages of continental drift, when the incipient fracture zone is a graben-type basin bounded by high angle faults on both sides.

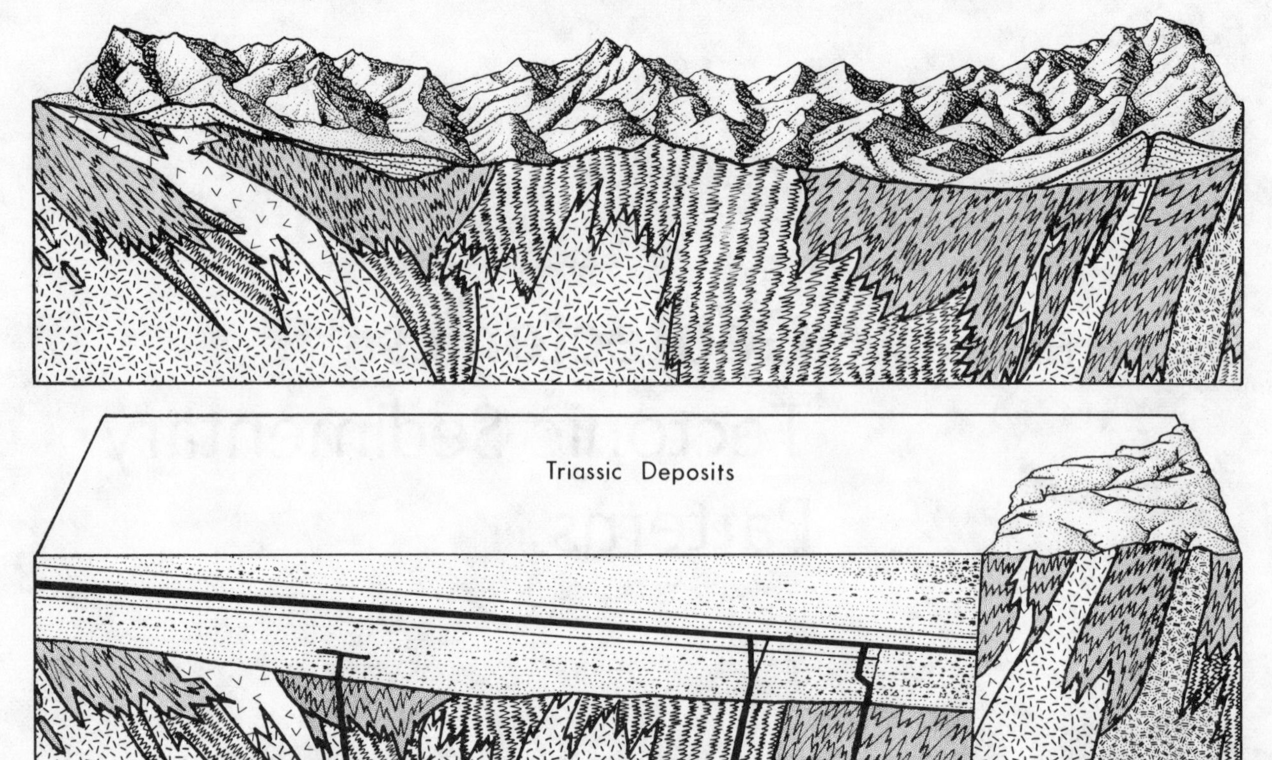

Figure 15.1 Development of a block fault sedimentary basin in the Triassic of the Connecticut Valley of New England.

Although alluvial fans may develop along any escarpment, or any place where a sediment-carrying stream experiences a sharp change in gradient (a *nickpoint*), they are most commonly found along the margins of high angle faults. For the geologist, ancient alluvial fans not only reveal the presence of ancient high angle faulting, but they may record the pulses of movement caused by intermittent slippage along the fault surface. The total thickness of the sediment also provides a measure of how much vertical displacement has been produced by faulting. Also, the sedimentary sequence carries, in upside down order, a complete record of the lithologic composition of the eroded away portions of the upthrown block, as deduced from pebbles and rock fragments in the sediment mass. No other depositional environment provides such a complete record of its source materials.

Alluvial fans may develop in both humid and arid climates, although modern ones are more pronounced in arid regions because of the lack of plant cover and slumping. Alluvial fan sediments are relatively porous, at least in the better sorted portions, and are often saturated with water. A high standing pile of water soaked sediment could easily experience very slow flowage with time, causing pebbles to scratch against each other and produce striated pavements if moving over bedrock. Thus it may be difficult to distinguish ancient bajadas from glacial tills; tills grade laterally into well-sorted outwash, just as the poorly sorted sediments at the fanhead grade laterally into better sorted sediments at the fan base (Figures 2.4, 2.5). Earthquakes triggered by movement along the bounding fault could cause rapid flowage of materials, or landslides.

The development of an alluvial fan takes place in several steps (Figure 15.2). First, upward displacement along the fault produces a steep escarpment; streams flowing across the fault

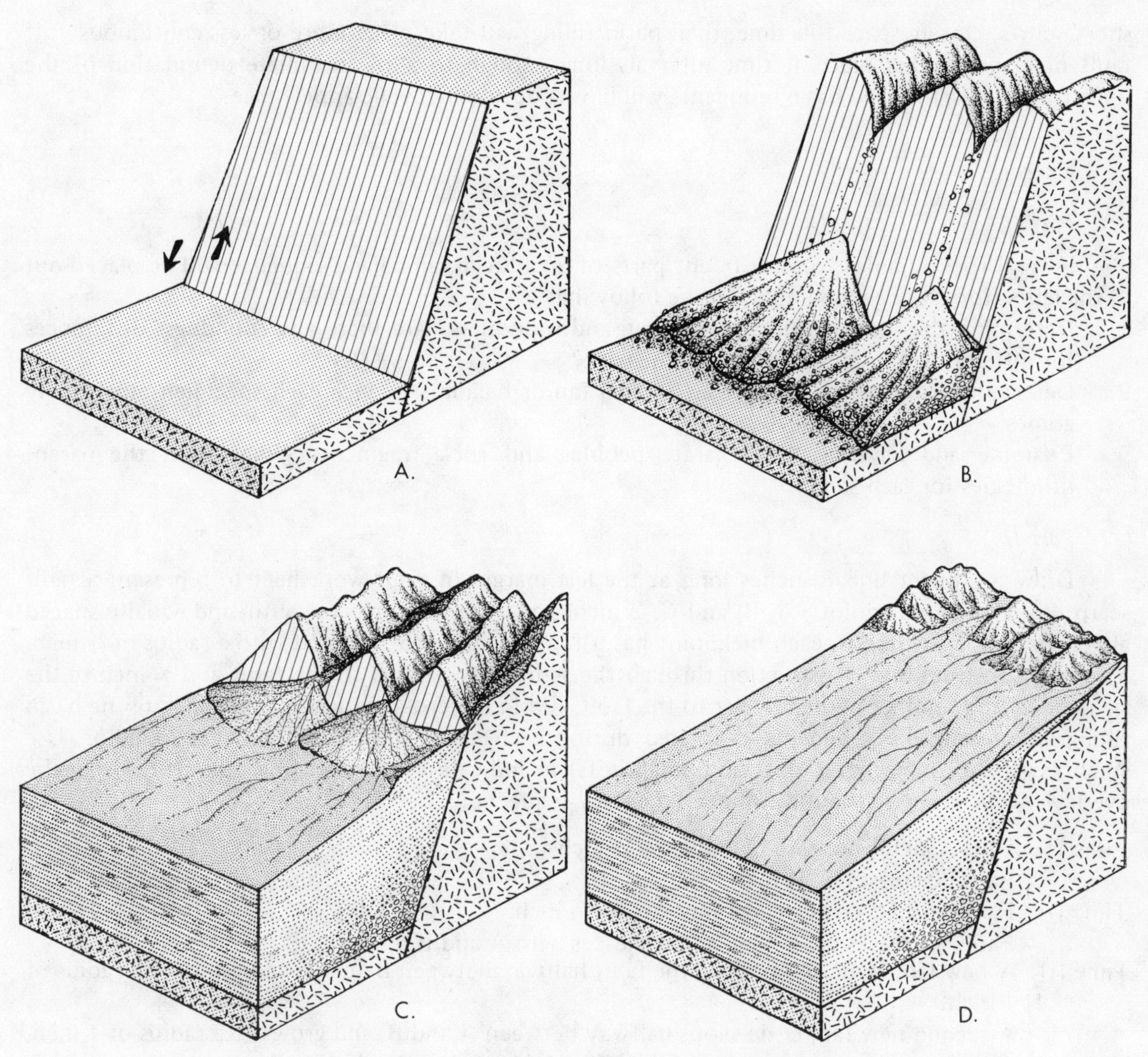

Figure 15.2 Stages in the development of alluvial fans

encounter a nickpoint, and dump their sediment load. The earliest sediments are the coarsest as well as the most poorly sorted. As aggradation proceeds, the sediment mass becomes higher and wider, and the source terrain on the upper block is lowered through erosion, thus gradually obliterating the nickpoint. The sediments should become finer and better sorted upwards in the fan, as well as from the fanhead to the base. In the final stage, all of the laterally adjacent fans along the scarp coalesce to form a bajada, a cover of sediment that completely fills in the basin.

In the process of filling a basin, fans may shift in position When two adjacent fans have been built up high enough, stream flow will shift to the low area in between the two old ones. Thus the development of a bajada basin-fill takes place as a series of *fill-ins*, very much like the development of a delta.

Once the basin has been partially or completely filled, renewed movement along the fault could start the whole process over again, with a new escarpment and new nickpoints, and generate another complete bajada sequence on top of the old one. The only limitation to the extent of such basin filling is that of the tectonic forces that control the faulting. If fault movement takes place in short spurts, closely spaced in time, then basin filling will take place more or less continuously. If fault movement takes place at time intervals long enough to allow complete denudation of the upthrown block, then the basin sediments will have a decidedly cyclic appearance.

Exercise

Part I

Several rock samples from different parts of an alluvial fan-bajada sequence will be placed out in the laboratory. For each sample, do the following:

1. Determine the modal grain size, sorting, and roundness. Are there any roundness differences for different grain sizes within the same rock?
2. Determine from which part of an alluvial fan or bajada sequence each specimen most likely comes.
3. Examine and identify the separate pebbles and rock fragments. What were the parent lithologies for each sample?

Part II

Draw a straight line 8 inches long at the left margin of your worksheet to represent a fault scarp. Place three nickpoints, A, B, and C, 2 inches apart from north to south, and equally spaced along the fault. At time 0, each nickpoint has a fan spreading to the right with a radius of ½ inch. You will construct one cross section through the sediments parallel to the fault and ¼ inch to the right of it, and another perpendicular to the fault. Fill in the cross section as you generate the basin filling. Represent the sediments deposited during each time interval by a different color. Use symbols (for example, large dots and small dots) to indicate coarse and fine sediment. Using the overlay sheets, generate the following sedimentary sequence:

Time I: Fans A and B increase in radius by ½ inch.
Fan C increases by ¼ inch.

Time II: Fans A and B coalesce and increase by ½ inch.
Fan C increases by ¼ inch, and coalesces with A and B.

Time III: A new fan, D, develops along the fault halfway between B and C, and grows to a radius of 1 inch.
A second new fan, E, develops halfway between A and B, and grows to a radius of 1 inch.

Time IV: All fans coalesce into a bajada that fills the basin, gradually thinning to the right, having an undulating border about 8 inches from the fault.

Time V: Renewed activity along the fault recreates fans with almost the same configuration as those at Time I.

Test Questions

1. Although both deltas and bajadas are generated by the process of fill-ins, what basic differences will there be in the three-dimensional geometry of the masses of sediment produced?
2. What criteria would enable you to distinguish a Precambrian tillite from an alluvial fan sediment?
3. Alluvial fans represent a direct effect of tectonic activity on the sedimentary record. In what way is deltaic sedimentation also controlled by tectonics? Could the abundance of detritals as opposed to carbonates be related to block uplifts or mountain building? Explain your answer.
4. Why are lava flows, dikes, and sills commonly associated with fault block basin sediments?
5. In arid climates, intermittent lakes often occur in the center of a fault block basin. What

chemical sediments and sedimentary structures might be formed in such lakes, and thus become interbedded in the alluvial fan sequence? What if they were large, extensive lakes with no external drainage?

6. What general relationships exist between time planes and units of homogeneous lithology in the alluvial fan sequence? Is Walther's Law applicable to these rocks?

Background Reading

Bull, W. B. 1972. Recognition of alluvial fan deposits in the stratigraphic record. In *Recognition of ancient sedimentary environments*, Society of Economic Paleontologists and Mineralogists, Spec. Pub. 16, p. 63-83.

Chapter Sixteen Biostratigraphy Patterns of Organic Evolution

Objectives

1. To examine the processes involved in the evolution of one biological species from another.
2. To examine the spatial patterns produced by transspecific evolution.
3. To determine the vertical and lateral distributions of species in evolutionary sequence preserved as skeletal fossils in a volume of sedimentary rock.

Important Terms

Species—The fundamental unit of the biological world, consisting of genetically similar organisms occupying a continuous range, maintaining a high level of functional similarity through either interbreeding or buffering by external environmental factors.

Population—A subgroup of a species; usually the organisms found in some easily defined segment of the total area occupied by the species.

Habitat—The area (or volume) that a species occupies.

Peripheral isolate—A population on the periphery of a species' range, which no longer maintains the genetic equilibrium of the parent species, through disruption of either interbreeding or external buffering systems.

Allopatric speciation—The rise of a new species from a peripheral isolate.

Sympatric speciation—The rise of a new species from populations *within* the range of the parent; generally an unusual or uncommon mode of speciation.

Phyletic speciation—The rise of a new species by gradual changes in *all* populations of the parent species; an intuitive possibility that has not yet been found in nature.

Saltation—The "jumping" or lack of intermediate forms from one species to another, either in space or time; caused by disruption of the genetic equilibrium (*homeostasis*) of the parent species, producing peripheral isolates that may differ greatly in morphology or behavior, but which do not necessarily differ greatly in genetic makeup. The newly speciated population will eventually attain its own homeostatic equilibrium.

Discussion

The process whereby new species arise from old ones to occupy or colonize new habitats is known as allopatric speciation. Sympatry, although possible, is relatively rare, and could result from invasion of the parental range by a daughter species that arose allopatrically. Normally, changing environmental conditions in the range of a species will not cause phyletic speciation, but will cause the range boundaries to migrate as environmental complexes shift position. Environmental changes that are global in scope are more likely to cause mass extinctions than phyletic evolution.

According to the concept of natural selection, species are environmentally adapted to the conditions in their habitat. Towards the periphery of their range, they are less well adapted, and drop to zero abundance at the range boundary. The sparse peripheral representatives of a species are subject to the greatest environmental stress; it is under these conditions that the genetic buffering systems may be disrupted to produce new forms.

Consider the complex mosaic of environmental conditions that would be found in the benthonic habitats along a broad continental shelf or an expansive inland sea. Differences in habitats could be due to distance from shore, nutrient supply from river mouths, sedimentational and substrate differences, salinity, photoperiod, depth, temperature, and wave and current activity. No two areas of any size could be identical for all environmental factors. It is thus expectable that different species occupy different parts of such large areas, and that they most likely arose by the allopatric process.

Exercise

Allopatric speciation in the marine environment could take place either parallel to the shoreline or in an onshore-offshore direction. This exercise will consider onshore-offshore speciation and onshore-offshore habitat migrations.

Draw on the worksheet a hypothetical shoreline at the top of the page, roughly parallel to the top edge of the paper. Label five locations, 1-5, going away from the shoreline, starting with the first location 1 inch from shore, and the successive sites evenly spaced 2 inches apart. Assume that the same sediment type is accumulating at each site, at more or less uniform rates. The evolutionary sequence will consist of six species, labeled A-G, with A as the common ancestor, and B-G as its descendants. Evolution will take place through five intervals of geologic time, labeled I-V, with I as the oldest and V the most recent. Assume that each species range (in space) will be 2 inches onshore-offshore, and 4 inches parallel to the shoreline. If the shoreline migrates, the species ranges will migrate with it, maintaining a constant distance from shore and a constant size. Using the overlay sheets, map the distribution of species through the following time sequence:

Time I: Species A appears, with habitat centered on locality 2.

Time II: The shoreline regresses to a position halfway between localities 1 and 2.
Species B arises shoreward of A, and C arises offshore of A.

Time III: The shoreline regresses to a position halfway between localities 2 and 3. No speciation occurs.

Time IV: The shoreline transgresses to a position halfway between localities 1 and 2.
Species A becomes extinct.
Species E arises from C, and invades the habitat formerly occupied by A.
Species D arises offshore from C.

Time V: The shoreline transgresses to its original (Time I) location.
Species C becomes extinct.
Species F arises offshore from E, and invades the habitat formerly occupied by C.
Species G arises offshore from D.

Construct a diagram showing this evolutionary pattern in time, with no reference to geographic location or stratigraphic position. Use time as a vertical axis, and plot the time span of each species as a vertical bar, with its base at the time the species appeared, and its top at the time of extinction. Show the ancestry of each species by connecting the base of its bar by a horizontal line to the bar of its parent (you should have placed parents and daughters next to each other). This diagram is a *phylogenetic tree*, and shows the true evolutionary relationships of these species in time.

Construct a stratigraphic column that represents the vertical thickness of rock deposited at each of the five localities. Assume that 1 inch (in reference to the map scale) of sediment accumulated at each locality during each time interval when the locality was under water. Draw each column 1 inch wide, and as high as the thickness of sediment that accumulated there. Show any surfaces of nondeposition by a wavy line. On the left of each column, indicate by brackets the time intervals represented in each part of the section. On the right of each column, indicate by brackets the species that are present as skeletal fossils in each part of the section.

How well does the vertical sequence of species at each locality represent the phylogenetic tree? Which localities show a true phylogenetic sequence? Which localities have a parent species overlying a daughter (a false phylogeny)? Which localities show displacement of one lineage by another? Which localities show stratigraphically recurrent species?

Test Questions

1. If you wished to trace a series of populations through geologic time that inhabited the offshore carbonate facies of a shoreline sequence, would it be more useful to study the vertical sequence of fossils at one locality, or the lateral sequence of fossil types at many localities? Which sampling plan would yield a greater duration of geologic time?
2. What general relationship exists between the distribution of allopatric species and time planes? Is the first appearance of a species in a local rock column a good indicator of a geologic time plane?
3. If a different sediment type had been accumulating at each species habitat, how would the resulting biostratigraphic patterns be altered? What types of biostratigraphic patterns would be found in a deltaic sequence?
4. Would planktonic marine microorganisms also show the same distributional patterns with respect to rock types? What factors might modify their distribution? Could the above exercise be applied to species of ancient pollen?
5. How useful is the principle of superposition used alone to determine evolutionary relationships?

Background Reading

Eldredge, N., and Gould, S. J., 1972, Punctuated equilibria: an alternative to phyletic gradualism, *in* Models in Paleobiology, Freeman, Cooper and Co., San Francisco, p. 82-115.

Smith, R. L., 1966, Ecology and Field Biology, Harper and Row, Publishers, New York, p. 448-84.

[illegible]

[illegible]

Questions

1. [illegible]
2. What general relationship exists between the distribution of [illegible] and time [illegible]
3. [illegible]

References cited

[illegible]

Chapter Seventeen

Geotectonics
Regional Sedimentary Patterns through Time

Objectives

1. To examine regional sedimentation across a large portion of the global surface through time.
2. To integrate tectonic activity, the distribution of modern environmental complexes, and detrital and carbonate sedimentation.
3. To determine the relationship of time planes to large-scale volumes of sedimentary rock.

Important Terms

Clastic wedge—A very thick fanlike cone of sediment deposited on the shelf area around a marine delta.

Geosyncline—A linear belt or structural trough characterized by great sediment thickness, usually in excess of 40,000 feet.

Orogeny—Tectonic uplift or mountain building event.

System—The body of sedimentary rocks deposited during each geologic time period; the volume of rock formed during the Ordovician Period is called the "Ordovician System." Systems may be further subdivided into *series* and *stage*, corresponding to the volumes of sediment formed during one epoch and age, respectively.

Discussion

Examine the map of the Gulf of Mexico region showing the location of different, contemporary sedimentational complexes (Figure 17.1). Along the coastline there are one major and three minor deltas, extensive stretches of detrital shoreline sedimentation, and one major carbonate area. In addition to shallow water sediments, there is a complementary series of deeper water sediments on the marine shelf. Extending from the subtidal areas of the Mississippi Delta to the outer edge of the continental shelf is a large cone or fanlike mass of detrital sediment, mostly marine clays and sands, derived from the river mouth. This mammoth, fanlike fill-in is very much like the clastic (detrital)wedges that dominated sedimentation along the Appalachian shoreline

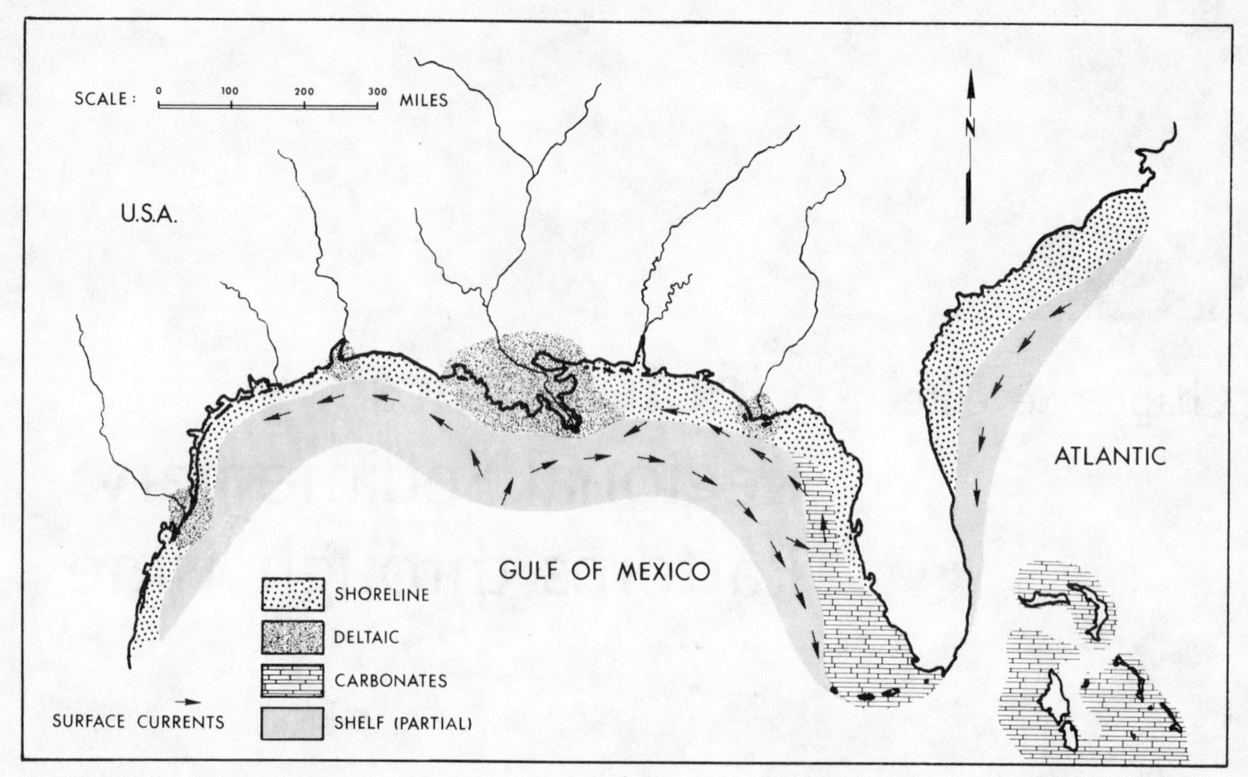

Figure 17.1 **Map of the Gulf of Mexico region, showing contemporary areas of carbonate, deltaic, shoreline, and shelf sedimentation.**

during the Paleozoic (Figure 17.2). Elsewhere, the marine shelf is blanketed with silt and clay where there is an adequate supply of detrital sediment; these deposits are much thinner and more sheetlike than the deltaic cone. Periodically, hurricanes cause very thin beds (only a few inches thick) of poorly sorted sand from the beaches and barrier islands to be spread out over the normal shelf muds. Where the supply of detritus is low, the shelf sediments are fine-grained carbonates (biomicrudites).

The Paleozoic sedimentary rocks of the eastern and central United States likewise reflect sedimentation in nearshore carbonate, shoreline, and deltaic areas, as well as offshore shelf sedimentation. Several major clastic wedges reflect the positions of major river mouths and their associated deltas. Elsewhere, coastal and offshore sedimentation analogous to modern sedimentation in the Gulf of Mexico was taking place. The positions of these clastic wedges shifted through time (middle Ordovician through Pennsylvanian), reflecting local tectonic uplifts that produced tremendous volumes of detrital sediment, generating a series of fill-ins that migrated up and down the coastline. The end result is a linear belt of detrital sediments (shales, siltstones, sandstones) of great thickness. Linear bands of great sediment thickness are sometimes termed geosynclines. Later tectonic events deformed these sedimentary rocks to produce the present Appalachian structure. Three of the major orogenic events in the Appalachian area have been named as follows:

Ordovician—Taconic Orogeny
Devonian—Acadian Orogeny
Permo-Carboniferous—Appalachian Orogeny.

The effects of these pulses of tectonic activity on sedimentation can be seen in Figure 17.2. The final tectonic event in the Permian, undoubtedly associated with continental separation, caused the

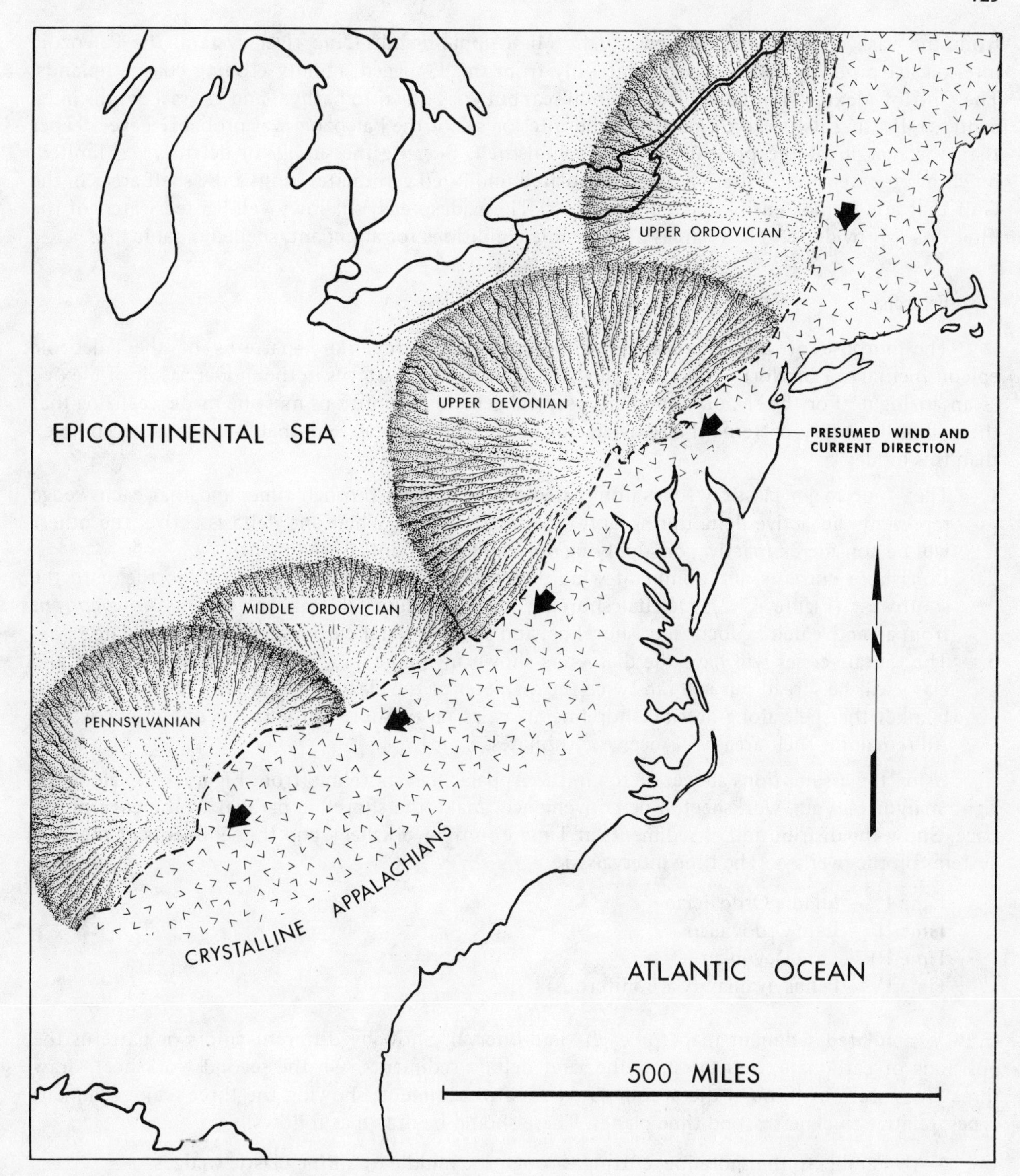

Figure 17.2 Map showing clastic wedges (deltas) of different ages in the Appalachian area, reflecting migrating tectonic uplift with time.

complete drainage (regression) of the epicontinental sea that had covered the continent during most of the Paleozoic, and the final abandonment of the Appalachian shoreline as thinner, deltaic fill-ins rapidly prograded westward.

Although some good analogies may be drawn between Paleozoic sedimentation and the modern Gulf of Mexico, several important differences must be noted. The Gulf receives sediment from the vast interior area drained by the Mississippi-Missouri-Ohio river system; the Paleozoic interior sea probably received sediment only from small, rugged, rapidly eroding coastal uplands. The Gulf of Mexico is not an epicontinental sea, but goes down to bathyal and abyssal depths in its center, reflecting its oceanic character. The interior sea of the Paleozoic was probably never deeper than the margin of the present-day continental shelf. Because the supply of detritus was limited, much more extensive offshore areas accumulated undiluted carbonates, unlike the shelf areas in the Gulf that are dominated by detrital sediments. The widespread, shallow, well-lighted waters of the interior sea provided extensive areas of favorable conditions for abundant, shelled organic life.

Exercise

The purpose of this exercise is to simulate the sedimentary patterns of the Paleozoic epicontinental sea of North America, using the sedimentary patterns in the modern Gulf of Mexico as an analogue. For the sake of simplicity, several general assumptions must be made, realizing that the real sedimentary patterns that existed during the Paleozoic were probably much more complex than this model:

1. That four major clastic wedges dominated sedimentation through time, and that each wedge represents an active delta during only one time interval. When one delta is active, the others will be considered inactive, or supplying only minor amounts of sediment.
2. Longshore currents and dominant wind directions are presumed to be from northeast to the southwest (Figure 17.2). Detrital shoreline sedimentation will take place only *downcurrent* from an active delta. Upcurrent, shoreline, and shallow water carbonates will accumulate.
3. The deltaic cones will have the diameters shown in Figure 17.2. Thin shelf detritals (silt and clay) will be spread out in a fan twice the diameter of the clastic wedge. Thin detritals will also blanket the shelf along detrital shoreline areas, as far offshore as the radius of a clastic wedge. All remaining shelf areas will receive carbonates.

Using the assumptions above, sketch in the Appalachian shoreline (from Figure 17.2) along the right margin of your worksheet. For convenience, make the shoreline parallel to the edge of the page. Show the distribution of sediments at Time I on the worksheet, and the sediments of the later systems on the overlays. The time intervals are:

Time I — Middle Ordovician
Time II — Late Ordovician
Time III — Late Devonian
Time IV — Pennsylvanian (Carboniferous)

Draw a simulated sediment map for each time interval. Show by different colors or patterns the positions of carbonate, detrital shoreline, and deltaic sediments. On the second worksheet, draw three cross sections through the accumulated mass of sediment, showing the three major sediment types, relative thicknesses, and time planes. These should be drawn as follows:

A-A′—Parallel to the shoreline, cutting through the middle of all the clastic wedges.
B-B′—Parallel to the shoreline, just seaward of the outer margins of the clastic wedges.
C-C′—Perpendicular to the shoreline, cutting through the middle of the Late Devonian wedge.

Test Questions

1. The Carboniferous of North America has been subdivided into a lower system, the Mississippian, and an upper system, the Pennsylvanian. The middle Mississippian rocks in the east central United States are largely undiluted carbonates. The upper Mississippian rocks are largely mixed detrital and carbonate shoreline and shallow water sediments. The Pennsylvanian

rocks are mostly deltaic and fluvial. How does this time-rock pattern compare to those generated above?

2. A common tactic in stratigraphy is to assume that the flooding of detritals over a carbonate area represents a regression, and that the development of carbonates on top of detritals represents a transgression. How valid is this approach? Can you produce carbonate-detrital oscillations in vertical rock sequences without changing the position of the shoreline?
3. What general relationships exist between regionally developed bodies of homogeneous lithology and time planes?
4. Considering the relative shallowness of the epicontinental sea, how is it possible that such a thick series of clastic wedges could have been built up along the Appalachian shoreline? What is the approximate ratio of presumed water depth to sediment thickness in this belt? What type of tectonic activity could have maintained this shoreline in a stable position for so long? Why is the Appalachian shoreline more nearly linear than the Gulf of Mexico shoreline? Why are volcanic rocks commonly interbedded with geosynclinal sediments? What is the relationship of the Applachian structure to the mid-Atlantic ridge?

Background Reading

King, P. B., 1959, The Evolution of North America, Princeton Univ. Press, Princeton, N.J., p. 41-88.

Shepard, F. P., Phleger, F. B., and Van Andel, T. H., 1960, Recent Sediments of the Northwest Gulf of Mexico, American Assoc. Petroleum Geologists, Tulsa, Oklahoma.

Sources of Illustrations

1.1 After Powers, 1953, fig. 1.
1.2 After Folk, 1965, p. 104, and Pettijohn, Potter and Siever, 1972, fig. A-1.
1.3 After Folk, 1965, p. 107.
1.4 Based on Krumbein and Sloss, 1963, figs. 4-3, 5-8.
2.1 Partially after Mason and Folk, 1958, figs. 1, 2.
2.2 From Hamblin and Howard, 1971, fig. 157.
2.3 Based on Fisk et al., 1954, figs. 1, 8, and Fisk, 1961 fig. 4.
2.4 From Hamblin and Howard, 1971, fig. 131.
2.5 Original.
3.1 After Margulis, 1970, frontispiece, and McAlester, 1968, p. 152.
3.2A After Fenton and Fenton, 1958, p. 76.
B Based on Moore, Lalicker and Fisher, 1952, fig. 4-12.
C-F Original.
G After Beerbower, 1968, fig. 10-16A.
3.3A After Glenn, 1904, Pl. 100, fig. 2.
B Original.
C After Nicholson and Lydekker, 1889, fig. 577.
D After Nicholson and Lydekker, 1889, fig. 559.
E After Easton, 1960, fig. 9.14, la.
F Original
G After Shrock and Twenhofel, 1953, fig. 10-41A.
H After Moore, Lalicker and Fisher, 1952, fig. 8-34, 12.
I After Martin, 1904, Pl. 52, fig. 5.
J Original.
3.4A Original.
B After Fenton and Fenton, 1958, p. 41.
C After Moore, Lalicker and Fisher, 1952, fig. 9-24.
D After Fenton and Fenton, 1958, p. 178.
E Original.
F After Easton, 1960, fig. 11.17, 8a.
G After Knight, 1856, fig. 2547.
H After Easton, 1960, fig. 11.31, 5a.
I Original.
J After Nicholson and Lydekker, 1889, fig. 782.
K-L Original.
3.5A After Easton, 1960, fig. 12.2.
B After Walcott, 1886, Pl. 16, fig. 6.
C After Walcott, 1910, Pl. 33, fig. 1.
D Original.
E After Whittington, 1959, fig. 353-1a.
F After Collinson, 1956, pl. 8.
G After Fenton and Fenton, 1958, p. 222.
H Original.
I After Clark and Twitchell, 1915, Pl. 14, figs. 3a, b.
J After Clark and Twitchell 1915 Pl. 43, fig 2a.
3.6A After Fenton and Fenton, 1958, p. 97.
B After Fenton and Fenton, 1958, p. 91.
C Original.
D After Ulrich, 1890, Pl. 63, fig. 8.
E After Moore, Lalicker and Fisher, 1952, fig. 5-4B, D.
F After Beerbower, 1968, fig. 11-3D.
G After Hall, 1852, Pl. 40D, fig. 1a.
H-I After Moore, Lalicker and Fisher, 1952, fig. 6-10.
J After Hall and Clarke, 1894, Pl. 56, figs. 14, 15.
K After Hall and Clarke, 1893, Pl. 8, fig. 17.
L After Hall and Clarke, 1894, Pl. 55, fig. 5.
M Original
3.7A After Shrock and Twenhofel, 1953, fig. 14-9B, C.
B-C After Fenton and Fenton, 1958, p. 128.
D After Roemer, 1860, Pl. III, fig. 2a.
E After Fenton and Fenton, 1958, p. 128.
F After Moore, Lalicker and Fisher, 1952, fig. 18-30, 20.
G After Wachsmuth and Springer, 1897, Pl. 9, fig. 1.
H After Springer, 1920, Pl. 53, fig. 41.
I After Fenton and Fenton, 1958, p. 130.
J Original.
K After Fenton and Fenton, 1958, p. 131.
L Original.
M-N After Fenton and Fenton, 1958, p. 130.
3.8 After House, 1971, fig. 14.1.
3.9 After Hedgpeth, 1966, fig. 5.
3.10-18 Original.
4.1 Original.
5.1 After Heckel, 1972, fig. 9.
5.2 After Heckel, 1972, fig. 3.
5.3A After Wells, 1954, Pl. 126, figs. 3, 5; Pl. 127, fig. 4; Pl. 128, fig. 3.
B Original.
C After Jones, 1907, figs. 155, 156.

5.4 Original.
5.5 After Bretsky, 1970, fig. 30.
5.6 After Schopf, 1969, fig. 1.
5.7 After Lowenstam, 1967, fig. 4.
6.1 Based on Lowenstam, 1950.
6.2 After Ehlers and Kesling, 1970, fig. 4.
6.3 Based on Bretsky, 1969.
7.1 After Folk, 1968, p. 157, 165.
7.2 After Anstey and Fowler, 1969, fig. 5.
7.3 After Folk and Robles, 1964, fig. 9.
7.4 After Shaw, 1964, figs. 8-1, 2.
7.5 After Ginsburg, 1956, fig. 9.
7.6A After Scoffin, 1970, fig. 18.
B After Scoffin, 1970, fig. 17.
C After Scoffin, 1970, fig. 8.
7.7 After Imbrie and Purdy, 1962, fig. 13.
8.1 Based on Hills, 1963, figs. 11-1, 2.
8.2 After Shrock, 1948, figs. 48, 49.
8.3 Original.
8.4 After Harms, 1969, fig. 15.
8.5 After Eicher, 1968, fig. 2-12.
8.6 After Shrock, 1948, fig. 207.
8.7 After Shrock, 1948, fig. 148.
8.8 After Shrock, 1948, fig. 100.
9.1 From Seilacher, 1967b, p. 74.
9.2A From Seilacher, 1967b, p. 75.
B-C Original.
D After Frey and Reineck, 1973, fig. 4.
9.3 Based on Seilacher, 1964, 1967a.
9.4 After Frey and Howard, 1969, fig. 2.
10.1 After Gebelein, 1969, fig. 14.
10.2 Based on Logan, Rezak and Ginsburg, 1964.
10.3 After Logan, 1961, fig. 3.
10.4 After Logan, 1961, fig. 2.
11.1 After Butler, 1969, figs. 1, 2.
11.2 Based on Hsu and Siegenthaler, 1969, and Shinn, Ginsburg, and Lloyd, 1965, fig. 4.
11.3 After Adams and Rhodes, 1960, fig. 3.
11.4 After Deffeyes, Lucia, and Weyl, 1965, fig. 15. (adaptation by Zenger, 1972, fig. 10).
11.5 After Thompson, 1970, fig. 18.
12.1 From Hamblin and Howard, 1971, fig. 160.
12.2 After Ferm, Milici, and Eason, 1972, fig. 2.
13.1A After Allen, 1964, fig. 1.
B-D After Fisher et al., 1969, fig. 37 (adapted from several sources).
13.2-3 Original.
13.4 After Fisher et al., 1969, fig. 13.
13.5 From Hamblin and Howard, 1971, fig. 210.
13.6 After Morgan, 1970, fig. 8.
13.7A After Coleman and Gagliano, 1964, fig. 5.
B After Coleman and Gagliano, 1964, fig. 1 (adapted from Kolb and Van Lopik, 1958).
13.8-10 Original.
14.1 After Strahler, 1966, fig. 5.
14.2 After Strahler, 1966, fig. 10.
14.3 From Hamblin and Howard, 1971, fig. 178.
14.4 From Hamblin and Howard, 1971, fig. 135.
14.5 After Dunbar and Waage, 1969, fig. 18-9.
14.6-9 Original.
15.1 After Moore, 1958, fig. 13.17.
15.2 Based in part on Bull, 1972, figs. 11, 14.
17.1 Based in part on Shepard et al., 1960.
17.2 After King, 1959, fig. 34.

Illustration Source References

Adams, J. E., and Rhodes, M. L., 1960, Dolomitization by seepage refluxion, American Assoc. Petroleum Geologists Bull. 44:1912-21.

Allen, J. R. L., 1964, Sedimentation in the modern delta of the River Niger, West Africa, in Van Straaten, L. M. J. U., ed., Deltaic and shallow marine deposits. Elsevier Publishing Co., Amsterdam, p. 26-34.

Anstey, R. L., and Fowler, M. L., 1969, Lithostratigraphy and depositional environment of the Eden Shale (Ordovician) in the tri-state area of Indiana, Kentucky and Ohio. J. Geol. 77:668-82.

Beerbower, J. R., 1968, Search for the past. Englewood Cliffs, N.J.: Prentice-Hall.

Bretsky, P. W., 1969, Central Appalachian Late Ordovician communities: Geol. Soc. Am. Bull. 80:193-212;

____. 1970. Upper Ordovician ecology of the central Appalachians: Yale Univ., Peabody Mus. Nat. History Bull., v. 34.

Bull, W. B., 1972, Recognition of alluvial fan deposits, in Rigby, J. K., and Hamblin, W. K., eds., Recognition of ancient sedimentary environments. Soc. Econ. Paleontologists and Mineralogists, Spec. Pub. 16, p. 63-83.

Butler, G. P., 1969, Modern evaporite deposition and geochemistry of coexisting brines, the Sabkha, Trucial Coast, Arabian Gulf: J. Sediment. Petrol. 39:70-89.

Clark, W. B., and Twitchell, M. W., 1915, The Mesozoic and Cenozoic Echinodermata of the United States. U.S. Geol. Survey, Monograph 54.

Coleman, J. M., and Gagliano, S. M., 1964, Cyclic sedimentation in the Mississippi River deltaic plain. Gulf Coast Assoc. Geol. Soc., Trans. 14:67-80.

Collinson, C. W., 1956, Guide for beginning fossil hunters. Illinois State Geol. Surv. Ed. Ser. 4.

Deffeyes, K. S.; Lucia, F. J.; and Weyl, P. K., 1965, Dolomitization of Recent and Plio-Pleistocene sediments by marine evaporite waters on Bonaire, Netherlands Antilles, in Pray, L. C., and Murray R. C., eds., Dolomitization and limestone diagenesis, Soc. Econ. Paleontologists and Mineralogists, Spec. Pub. 13, p. 71-87.

Dunbar, C. O., and Waage, K. M., 1969, Historical geology, John Wiley and Sons, New York.

Easton, W. H., 1960, Invertebrate paleontology, Harper and Bros., Publishers, New York.

Ehlers, G. M., and Kesling, R. V., 1970, Devonian strata of Alpena and Presque Isle Counties, Michigan: Michigan Basin Geol. Soc., Guide Book for field trips.

Eicher, D. L., 1968, Geologic Time, Prentice-Hall, Englewood Cliffs, N.J.

Fenton, C. L., and Fenton, M. A., 1958, The fossil book, Doubleday and Co., Garden City, N.Y.

Ferm, J. C., Milici, R. C., and Eason, J. E., 1972, Carboniferous depositional environments in the Cumberland Plateau of southern Tennessee and northern Alabama; Tennessee Div. Geology, Rept. Inv. 33.

Fisher, W. L., Brown, L. F., Scott, A. J., and McGowen, J. H., 1969, Delta systems in the exploration for oil and gas: Bur. Economic Geology, Univ. Texas at Austin.

Fisk, H. N., 1961, Bar finger sands of the Mississippi Delta, in Peterson, J. A., and Osmond, J. C., eds., Geometry of sandstone bodies, American Assoc. Petroleum Geologists, Tulsa, Okla., p. 29-52.

____, McFarlan, E., Kolb, C. R., and Wilbert, L. J., 1954, Sedimentary framework of the modern Mississippi Delta: J. Sedimentary Petrology, v. 24, p. 76-99.

Folk, R. L., 1965, Petrology of sedimentary rocks, Hemphill's Austin, Texas.

Folk, R. L., and Robles, R., 1964, Carbonate sands of Isla Perez, Alacran reef complex, Yucatan: J. Geology, v. 72. p. 255-92.

Frey, R. W., and Howard, J. D., 1969, A profile of biogenic sedimentary structures in a Holocene barrier island-salt marsh complex, Georgia: Gulf Coast Association of Geological Societies, Transactions, v. 19, p. 427.

Frey, R. W., and Reineck, H. E., 1973, Holocene sediments of the Georgia coastal area, in Frey, R. W., ed., The Neogene of the Georgia coast, Dept. of Geology, Univ. of Georgia, Athens, p. 1-58.

Gebelein, C. D., 1969, Distribution, morphology, and accretion rate of Recent subtidal algal stromatolites, Bermuda: J. Sedimentary Petrology, v. 39, p. 49-69.

Ginsburg, R. N., 1956, Environmental relationships of grain size and constituent particles in some south Florida car-

bonate sediments: American Assoc. Petroleum Geologists Bull, v. 40, p. 2384-2427.
Glenn, L. C., 1904, Pelecypoda, in Miocene, Maryland Geological Survey, Baltimore, p. 274-401.
Hall, J., 1852, Paleontology of New York, v. 2, C. Van Benthuysen, Albany. N.Y.
_____, and Clarke, J. M., 1893, Paleontology of New York, v. 8, pt. 1, C. Van Benthuysen, Albany, N.Y.
_____, and _____, 1894, Paleontology of New York, v. 8, pt. 2, C. Van Benthuysen, Albany, N.Y.
Hamblin, W. K., and Howard, J. D., 1971, Physical geology laboratory manual, Burgess Publishing Co., Minneapolis, Minn.
Harms, J. C., 1967, Hydraulic significance of some sand ripples: Geol. Soc. America, Bull., v. 80, P. 363-96.
Hills, E. S., 1963, Elements of structural geology, John Wiley and Sons, New York.
Heckel, P. H., 1972, Recognition of ancient shallow marine environments, in Rigby, J. K., and Hamblin, W. K., eds., Recognition of ancient sedimentary environments: Soc. Economic Paleontologists and Mineralogists, Spec. Pub. 16, p. 226-76.
Hedgpeth, J. W., 1966, Marine ecology, in Fairbridge, R. W., ed., The encyclopedia of oceanography, Reinhold Publishing Co., New York, p. 454-59.
House, M. R., 1971, Evolution and the fossil records, in Gass, I. G., and Smith, P. J., eds., Understanding the earth, M: I. T. Press, Cambridge, Mass., p. 192-211.
Hsu, K. J., and Siegenthaler, C., 1969, Preliminary experiments on hydrodynamic movement induced by evaporation and their bearing on the dolomite problem: Sedimentalogy, v. 12, p. 11-25.
Imbrie, J., and Purdy, E. G., 1962, Classification of modern Bahamian carbonate sediments, in Ham, W. E., ed., Classification of carbonate rocks: American Assoc. Petroleum Geologists, Mem. 1, p. 253-72.
Jones, F. W., 1907, On the growth forms and supposed species in corals: Zool. Soc. London, Proc., p. 518-56.
King, P. B., 1959, The evolution of North America, Princeton Univ. Press.
Knight, C., 1856-1858, Reptiles, fishes, mollusca, and insects, v. 2, Pictorial museum of animated nature, London Printing and Publishing Co. Ltd., New York.
Kolb, C. R., and Van Lopik, J. R., 1958, Geology of the Mississippi River deltaic plain, southeastern Louisiana: U.S. Army Corps of Engineers, Waterways Experiment Station Tech. Rep. 3-483 and 3-484.
Krumbein, W. C., and Sloss, L. L., 1963, Stratigraphy and sedimentation, W. H. Freeman and Co., San Francisco.
Logan, B. W., 1961, *Cryptozoon* and associate stromatolites from the Recent, Shark Bay, Western Australia: J. Geology, v. 69, p. 517-33.
Logan B. W., Rezak, R., and Ginsburg, R. N., 1964, Classification and environmental significance of algal stromatolites: J. Geology, v. 72, p. 68-83.
Lowenstam, H. A., 1950, Niagaran reefs of the Great Lakes area: J. Geology, v. 58, p. 430-87.
_____, 1967, Adaptive traits in skeletal morphology, American Geol. Inst., Short Course Lecture Notes, Paleoecology, p. HL1-HL3.
McAlester, A. L., 1968, The history of life, Prentice-Hall, Englewood Cliffs, N. J.
McChesney, J. H., 1868, Descriptions of fossils from the Paleozoic rocks of the western United States, with illustrations: Chicago Acad. Sciences, Trans., Art 1, p. 1-57.
Margulis, L., 1970, Origin of eukaryotic cells, Yale Univ. Press, New Haven, Conn.
Martin, G. C., 1904, Gastropoda, in Miocene, Maryland Geological Survey, Baltimore, p. 131-269.
Mason, C. C., and Folk, R. L., 1958, Differentiation of beach, dune, and aeolian flat environments by size analysis, Mustang Island, Texas: J. Sedimentary Petrology, v. 28, p. 211-16.
Moore, R. C., 1958, Introduction to historical geology, McGraw-Hill Book Co., New York.
Moore, R. C., Lalicker, C. G., and Fischer, A. G,, 1952, Invertebrate fossils, McGraw-Hill Book Co., New York
Morgan, J. P., 1970, Deltas – a resumé: J. Geol. Education, v. 18, p. 107-117.
Nicholson, H. A., and Lydekker, R., 1889, Manual of paleontology, William Blackwood and Sons, Edinburgh and London.
Pettijohn, F. J., Potter, P. E., and Siever, R., 1972, Sand and sandstone, Springer Verlag, New York.
Powers, M. C., 1953, A new roundness scale for sedimentary particles: J. Sedimentary Petrology, v. 23, p. 117-19.
Roemer, F., 1860, Die silurischen fauna des westlichen Tennessee, Eduard Trewendt, Breslau.
Schopf, T. J. M., 1969, Paleoecology of ectoprocts (bryozoans): J. Paleontology, v. 43, p. 234-44.
Scoffin, T. P., 1970, The trapping and binding of subtidal carbonate sediments by marine vegetation in Bimini Lagoon, Bahamas: J. Sedimentary Petrology, v. 40, p. 249-73.
Seilacher, Adolf, 1964, Biogenic sedimentary structures, in Imbrie, J., and Newall, N., eds., Approaches to paleoecology, John Wiley and Sons, New York, p. 296-316.
_____, 1967a, Fossil behavior: Scientific American, v. 217, p. 72-80.
_____, 1967b, Bathymetry of trace fossils: Marine Geology, v. 5, p. 413-28.
Shaw, A. B., 1964, Time in stratigraphy, McGraw-Hill Book Co., New York.
Shepard, F. P., Phleger, F. B., and Van Andel, T. H., 1960, Recent sediments of the northwest Gulf of Mexico, American Assoc. Petroleum Geologists, Tulsa., Okla.
Shinn, E. A., Lloyd, R. M., and Ginsburg, R. N., 1965, Recent supratidal dolomite from Andros Island, Bahamas, in Pray, L. C., and Murray, R. C., eds., Dolomitization and limestone diagenesis: Soc. Economic Paleontologists and Mineralogists, Spec. Pub. 13, p. 112-23.
Shrock, R. R., 1948, Sequence in layered rocks, McGraw-Hill Book Co., New York.
Shrock, R. R. and Twenhofel, W. H., 1953, Principles of invertebrate paleontology, McGraw-Hill Book Co., New York.
Springer, F., 1920, Crinoidea Flexibilia: Smithsonian Inst., Pub. 2501, p. 1-486.
Strahler, A. N., 1966, A geologist's view of Cape Cod, The Natural History Press, Garden City, N.Y.
Thompson, A. M., 1970, Tidal flat deposition and early dolomitization in Upper Ordovocian rocks of southern Appalachian valley and ridge: J. Sedimentary Petrology, v. 40, p. 1271-86.
Ulrich, E. O., 1890, Paleozoic Bryozoa, in Paleontology of Illinois: Illinois Geol. Survey, Geology and Paleontology, v. 8, pt. 2, sect. 6, p. 285-688.
Wachsmuth, C., and Springer, F., 1897, North American Crinoidea Camerata: Mus. Comparative Zoology (Harvard), v. M20, 21, p. 10897.
Walcott, C. D., 1886, Second contribution to the studies on the Cambrian faunas of North America: U.S. Geol. Survey, Bull. 30.
Walcott, C. D., 1910, *Olenellus* and other genera of Mesohacidae, in Cambrian geology and paleontology, no. 6: Smithosonian Misc. Coll. n. 1805, v. 53, p. 231-422.
Wells, J. W., 1954, Recent corals of the Marshall Islands: U.S. Geol. Survey, Professional Paper 260-I, p. IV, 385-486.
Whittington, H. B., 1959, Trilobita, in Harrington, H. J., et al., Arthropoda 1, Part O, Treatise on Invertebrate Paleontology, Geol. Society America and Univ. Kansas Press.
Zenger, D. H., 1972, Dolomitization and uniformitarianism: J. Geol. Education, v. 20, p. 107-124.

Worksheet

Worksheet

Worksheet

Worksheet

Worksheet

Worksheet

Worksheet